Ocular Hypertension - The Knowns and Unknowns

Edited by Michele Lanza

Published in London, United Kingdom

IntechOpen

Supporting open minds since 2005

Ocular Hypertension - The Knowns and Unknowns
http://dx.doi.org/10.5772/intechopen.91529
Edited by Michele Lanza

Contributors
Ashutosh Dayal, Sonal P. Yadav, Sayantan Biswas, Jan Lestak, Martin Fůs, Daniel Laroche Laroche, Kara Dorothy Rickford, Luis Nino-de-Rivera, Diego Cervera, Paola Castillo-Juarez, John Chen, Fidan Jmor

First published in London, United Kingdom, 2021 by IntechOpen
IntechOpen is the global imprint of INTECHOPEN LIMITED, registered in England and Wales, registration number: 11086078, 5 Princes Gate Court, London, SW7 2QJ, United Kingdom
Printed in Croatia

British Library Cataloguing-in-Publication Data
A catalogue record for this book is available from the British Library

Additional hard and PDF copies can be obtained from orders@intechopen.com

Ocular Hypertension - The Knowns and Unknowns
Edited by Michele Lanza
p. cm.
Print ISBN 978-1-83969-337-3
Online ISBN 978-1-83969-338-0
eBook (PDF) ISBN 978-1-83969-339-7

Meet the editor

Michele Lanza graduated in Medicine and Surgery at Medical School of Seconda Università di Napoli, Italy, after which he began a residency program in ophthalmology in 2001. Today he is Associate Professor of Ophthalmology at Università della Campania, Luigi Vanvitelli, Italy. His fields of interest are anterior segment disease, keratoconus, glaucoma, corneal dystrophies, and cataract. His research topics include intraocular lens (IOL) power calculation, eye modification induced by refractive surgery, glaucoma progression, and validation of new diagnostic devices in ophthalmology. Dr. Lanza is the author of more than 100 papers published in international and Italian scientific journals and book chapters and is the editor of two international books. He also has more than 150 communications or posters in the most important international and Italian ophthalmology conferences to his credit.

Contents

Preface

This book provides useful information to physicians who manage patients with ocular hypertension. It answers questions such as, is ocular hypertension an actual disease? How it is possible to detect it? How does this condition force people to modify their lifestyles? Because the optic nerve damage due to ocular hypertension is one of the most important causes of irreversible vision loss, it is essential to manage this condition in the best way possible. The book examines advances in technology that allow for earlier diagnosis and better management of patients, but physicians must be aware of the limits and challenges of new devices recently introduced. Finally, this volume helps physicians improve their abilities to manage ocular hypertension and its connections with other ocular and systemic diseases.

Michele Lanza
University of Campania "Luigi Vanvitelli",
Italy

Section 1

Ocular Hypertension

Chapter 1

Ocular Hypertension in Blacks

Daniel Laroche and Kara Rickford

Abstract

Ocular hypertension occurs when intraocular pressure (IOP) is greater than the normal range with no evidence of vision loss or damage to the optic nerve. Individuals with ocular hypertension have an increased risk for glaucoma. The mean normal IOP is 15 mmHg and the mean IOP of untreated glaucoma is 18 mmHg. Elevated IOP commonly occurs in patients over the age of 50 and is often due to enlargement of the lens, narrowing of the angle, iridolenticular apposition, and pigment liberation that obstructs the trabecular meshwork. Cataract surgery and lensectomy can lower IOP and reduce the risk of glaucoma. The global wealth inequality of Blacks has created health inequities that have led to decreased access to surgical care contributing to higher rates of blindness from glaucoma. Greater education on the benefits of early cataract surgery and trabecular bypass for higher risk patients, as well as addressing wealth and health inequities, can help to bend the curve of blindness from glaucoma.

Keywords: ocular hypertension, glaucoma, Blacks, cataract surgery, trabecular bypass, African-American

1. Introduction

Ocular hypertension occurs when the pressure in the eye (intraocular pressure or IOP) is beyond the normal value with no signs of vision loss or damage to the optic nerve [1]. With ocular hypertension, the aqueous humor (fluid produced by the eye) is poorly drained. The buildup of fluid in the eye leads to an increase in IOP that could potentially lead to damage of the optic nerve, causing glaucoma [2]. The mean normal IOP is 15 mmHg and the mean IOP of untreated glaucoma is 18 mm Hg [1]. Ocular hypertension typically presents with no signs or symptoms, making it difficult to detect without access to an eye exam. Individuals with elevated IOP may be treated with cataract surgery and lensectomy [2, 3]. To properly address populations at risk for ocular hypertension, it is advantageous to determine how demographic variables may impact an individual's susceptibility to blindness. Demographic variables are innate or non-changeable determinants of a disease. Addressing inequities in wealth, health, and access to medical care, as well as improved education on the benefits of early surgical intervention, can bend the curve of blindness from glaucoma. In this chapter, we use epidemiologic studies focusing specifically on Blacks to describe the prevalence and management of ocular hypertension.

2. Prevalence of ocular hypertension in blacks

In 2019, Black Americans made up 12.8% of US population, accounting for over 42 million people [4]. Although Blacks make up a minority of the population, many eye diseases, including ocular hypertension and glaucoma, affect a disproportionate number of Blacks, leading to higher rates of vision loss than documented in white-Americans [5]. Definitions of race and ethnicity have been ill-defined in past medical literature, with many overlaps. Therefore, the term "Blacks" in this context refers to an individual of black African descent. The population of Blacks in the Caribbean is over 21 million and in Africa is close to one billion [6]. There are also issues of decreased access to surgery in both locations [7].

While it has been universally accepted and documented that Blacks have higher prevalence of ocular hypertension, the degree of prevalence may differ for varying black populations. For example, the black-American and black-Caribbean populations studied in the Baltimore Eye Survey and the Barbados Eye Study, respectively, are ethnically unique. Both populations of Blacks presented with a high prevalence of ocular hypertension, but to a different degree. The prevalence of ocular hypertension in the black-Caribbean population was reported at levels nearly twice that of the black-American population [8–11]. Studies have also reported a notably higher prevalence of ocular hypertension in Blacks in comparison to other racial groups (primarily white) [12, 13].

In response to a lack of substantial ocular research with Black study participants, extensive population-based studies including the Baltimore Eye Survey [9], The Ocular Hypertension Treatment Study [2], The Barbados Eye Study [8], and the African American Eye Disease Study [13] were created to address the disproportionate prevalence of eye diseases present in Blacks. Further studies are needed to continue to build upon this body of research, particularly to look at earlier interventions of cataract surgery and trabecular bypass as an earlier intervention to prevent glaucoma.

3. Mechanism of ocular hypertension in blacks

Studies have shown that with age the crystalline lens increases in width. During accommodation, the iris bows posteriorly. With age there is increased contact between the posterior iris pigment epithelium and lens zonules leading to pigment liberation and obstruction of the trabecular meshwork [14]. This is often seen with heavier pigment in the trabecular meshwork inferiorly compared to superiorly on gonioscopy [15]. The increased width of the lens can also lead to pupillary block and iris obstruction of the trabecular meshwork leading to elevated intraocular pressure. This common mechanism of ocular hypertension in persons over the age of 50 is often overlooked by physicians. Current physicians and those in training must be better educated to look for this clinically and intervene promptly. Early cataract surgery and lensectomy is beneficial to remove the large lens and trabecular bypass to restore aqueous outflow via the obstructed trabecular meshwork [14].

4. Genetic influence on ocular hypertension

Previous studies have shown intraocular pressure (IOP) to be highly heritable, indicating possible genetic influence on the development of ocular hypertension [12, 16]. There is additional substantial evidence suggesting that ocular hypertension leading to glaucoma may have a genetic component [17], but the specific genetic risk factors have not yet been identified. A 2012 genome wide association study conducted in 11,972 participants from The Netherlands, UK, Australia, and Canada investigated candidate genes in human ocular tissue to identify susceptibility to elevated IOP and glaucoma [12]. Elevated IOP commonly occurs in patients over the age of 50 and often presents with enlargement of the lens, narrowing of the angle, iridolenticular apposition, or pigment liberation that obstructs the trabecular meshwork. Genes regulating these ocular components were studied and the results revealed that genetic variants expressed in genes GAS7 and TMCO1 were associated with changes in IOP in the populations studied. Both revealed only marginal evidence for ocular hypertension, as GAS7 was associated with a 0.19 mmHg decrease in IOP and TMCO1 was associated with a 0.28 mmHg increase in IOP [12]. Additional findings revealed that individuals of European ancestry expressed the GAS7 variant at 0.44 frequency while those of African ancestry expressed the same variant at 0.12 frequency [12]. The lower frequency of this variant in Blacks may reflect the elevated IOP common in individuals of African descent and requires further research.

While impressive strives have been made over the past two decades to identify genetic components of ocular diseases [18], a comprehensive understanding of the pathophysiology has frequently been limited to individuals of European and Asian ancestry, requiring an increased need for genetic research in Blacks and other understudied populations. For example, multiple genetic variants in genes associated with elevated IOP were discovered in non-Black populations and a majority do not replicate, nor have an effect, in Blacks [19–21]. In response to an increased need for the identification of genetic risk factors that underlie elevated IOP in the understudied population of Blacks, the Primary Open-Angle African American Glaucoma Genetics (POAAGG) study was created in 2014 and took place over the course of five years to address these research disparities [22]. This study identified a genetic variation known as a single nucleotide polymorphism (SNP) involved in the homeostasis of the trabecular meshwork [23]. The trabecular meshwork (TM) is located in the anterior portion of the eye and regulates the outflow of the aqueous humor into circulation [24]. If resistance increases in the TM during aqueous humor outflow, intraocular pressure may rise leading to ocular hypertension. By identifying a genetic variant that may affect the TM in Blacks, the POAAGG study has made a pertinent finding to our understanding of the role of genetics in ocular hypertension and glaucoma. As one of the first large cohort studies with over 5,000 study participants, additional analyses are needed to further validate the implications of this study.

In addition, the progression of elevated IOP in Blacks leading to ocular hypertension is likely a combination of genetic, environmental, aging and socioeconomic factors, as well as others not mentioned. These demographic variables will continue to be explored throughout this chapter.

4.1 Central corneal thickness in ocular hypertension

Intraocular pressure is routinely measured in clinical practice to assess various conditions within the eye, including that of the optic nerve and visual field [25]. Goldmann applanation tonometry is the most common technique used to measure IOP, but its accuracy and use as a diagnostic tool may be impeded by the rigidity of the cornea [25]. A thicker cornea may cause an overestimate of IOP and a thinner cornea may cause an underestimate of IOP. The consensus on the necessity to correct IOP based on central corneal thickness is not yet clear. While CCT is statistically significant as a predictor of glaucoma development [2], it does not present as an independent risk factor [26].

The Ocular Hypertension Treatment Study (OHTS) and the European Glaucoma Prevention Study (EGPS) recognized central corneal thickness as one of the most significant predictors for primary open-angle glaucoma [2, 27]. The mean central corneal thickness is about 560 μ m and the risk for developing ocular hypertension has been reported to nearly double (hazard ratio of 1.82) for every 40 μm decrease [28]. Patients with thin corneas (<555 μm) [2] may present with an underestimated IOP reading, placing the individual at potential risk if actual IOP is elevated. The primary diagnostic criteria for ocular hypertension is IOP, so any factor that hinders this measurement may lead to an errant diagnosis. Patients with ocular hypertension typically present with thicker corneas, which may lead to an overestimation of IOP, while primary open angle glaucoma patients present with thinner corneas [29]. While the influence of elevated IOP on central corneal thickness has not yet been determined, individuals whose IOPs have been reduced pharmacologically by at least 20% demonstrated no change in corneal thickness [30].

Differences in central corneal thickness were noted between black Americans and white Americans. In the OHTS, Blacks were found to have thinner central corneal thickness (555.7 μm), resulting in lower applanation readings and a miscalculated estimation of the true level of IOP [30]. The South African Eye Study [31] also measured differences in central corneal thickness and compared the findings to measurements of intraocular pressure in Blacks, mixed ethnicity peoples, and whites. The findings revealed that Blacks had the thinnest corneas and highest IOP, followed by mixed ethnicity then white individuals.

These results suggest the possible need for refining the risk factor definitions when measuring central corneal thickness and IOP in varying populations. While obtaining a central corneal thickness measurements for all patients may not be necessary, patients with ocular hypertension should continue to be monitored to measure accurate IOP and determine possible susceptibility to glaucoma.

4.2 Morphological changes in the retinal nerve fiber layer in ocular hypertension

Differing from glaucoma, ocular hypertension presents with a normal optic nerve and no signs of damage. Ocular hypertension is often a precursor to glaucoma as abnormally high pressures in the eye may lead to damage of the optic nerve causing vision loss or blindness [1]. Studies have indicated differences in the structure of the optic nerve between Blacks and whites [32–33]. The optic disc area was 12% larger in Blacks compared to Whites [32]. The larger optic nerve may cause a greater strain at similar pressure levels, but it is not clear if larger optic discs affect one's susceptibility to ocular hypertension as there are incongruous reports [10, 34]. The impact of these differences has been postulated to affect the increased susceptibility of Blacks to ocular hypertension and glaucoma.

The retinal nerve fiber layer (RNFL) is primarily comprised of retinal ganglion cell axons that progressively diminish in glaucoma. As a result, the RNFL thins considerably and may present as an early manifestation of glaucoma [35]. As a precursor to glaucoma, RNFL was measured in patients with ocular hypertension and the results revealed a significant thinning of RNFL of about 15% in ocular hypertensive eyes as compared to normal eyes [36]. Other studies have yet to demonstrate significant differences RNFL between normal eyes and those with ocular hypertension, possibly due to the sensitivity of the instruments used to measure and the study population [36, 37].

5. Inequities contributing to ocular hypertension in blacks

Vision loss is a pertinent public health challenge that requires the efforts of many to overcome [38, 39]. Addressing these disparities involves contending with the pervasive economic and racial inequalities that have had a disproportionate impact on Blacks, particularly in healthcare utilization. These inequities are evident in a 2020 study documenting the recency of eye examinations among black adults over the age of 55 [40]. In this study, 13.4% of participants (n = 740) reported having no eye examination in the last five years and nearly 25% had not had an eye exam in the last year [40]. Concerningly, 20% of study participants with diabetes mellitus were not instructed by other healthcare providers to seek annual eye examinations.

5.1 Health and wealth inequities in blacks

Systemic and social inequities have resulted in poor health outcomes in Blacks [41]. When examining wealth in the United States, there is countless evidence of extensive racial disparities. In 2016, the net worth of the average white family in the US was nearly ten times more than that of a Black family at $171,000 and $17,150 respectively [42]. These extensive differences in wealth and income reflect the consequences of years of discrimination, segregation, and inequality that mark the history of the US from its inception. The wealth gap between Blacks and whites in the US demonstrates the differences in opportunity afforded to citizens [42]. Colonialism has contributed to similar wealth disparities in the Caribbean and Africa. Differing from the circumstances in developing countries, the eye health care system in the United States is highly capable of delivering the care necessary to treat patients [41, 43]. However, much improvement is needed in the means by which education is delivered to the public and effective screening may take place.

The history of medicine and health care in the United States is tainted by a myriad of forms of injustice and violence towards Blacks that includes segregation of medical facilities, unequal healthcare access, and disdainful medical experimentation [44, 45]. Today, these inequalities are especially evident in employment, housing, and wealth opportunities in medically underserved areas and populations (MUA/P) [46–48]. MUA/P have been defined by the Health Resources and Services Administration as areas or populations having too few health care providers, high poverty or high elderly populations [49]. In addition, there are also social factors that have had strong implications on the health outcomes of Blacks, particularly poverty, food insecurity, and affordable housing. Low-socioeconomic status and race have been independently associated with increased vision loss placing poor Blacks at an increased risk [47]. These social factors that have often led to poor health outcomes in Blacks are rooted in racism and implicit biases that have to be recognized and changed at the personal, medical, and institutional level in order to lead to change [50].

Many studies have reported the association between visual impairment and poor quality of life, as well as physical and mental illness [51–53]. Unilateral and bilateral vision loss and blindness can impact a person's quality of life by affecting their ability to read, walk, commute, and carry out daily activities [54]. In addition to the disparities previously mentioned, blindness can exasperate the inequities faced by Blacks in the US. Early treatment of ocular hypertension by reducing elevated IOP by 20% can reduce the risk of developing glaucoma in half [2], thereby reducing the risk of blindness. Earlier cataract surgery, clear lensectomy, and trabecular bypass may reduce it even more. Implementing measures to address ocular hypertension in Blacks can help reduce the risk of blindness and address health inequities in the medical community. In addition, public policy is needed to develop models of healthcare that make services more accessible, particularly in communities that are medically underserved.

5.2 Insurance and access to care

Access to health care can impact one's health outcomes. The utilization of healthcare may be determined by whether people know care is needed, whether obtaining care is wanted, and whether care can be accessed [55]. Access is often used to describe the ease of obtaining care, including its availability, the accommodations provided, and affordability. Health care in the United States often cannot be utilized without insurance, regardless of the presence of a healthcare provider that is geographically accessible. The public health challenge regarding ocular hypertension is that if the elevated IOP was detected earlier on, further exasperation of the condition could be slowed and potential diseases could be prevented [56, 57]. With newer surgical approaches progression can be halted with earlier cataract surgery/ clear lensectomy and trabecular bypass.

Successful treatments for elevated IOP have included topical medications, surgery, or laser [58]. Reducing IOP significantly may lead to a delay in progression to optic nerve damage, visual field loss, or glaucoma [59]. Several studies have reported the impact of lack of medical care on health outcomes [60, 61]. The Salisbury Eye Evaluation Study [62] was a population-based study that sought to investigate the causes of blindness and visual impairments of adults between the ages of 65 and 84. The study revealed higher levels of blindness and visual impairments in Blacks compared to whites, with 37% of the conditions classified as surgically treatable and 44% categorized as targets for low vision remediation. The study was not able to identify patients whose eye condition was amenable and chose not to undergo surgery for reasons including financial barriers, fear of the surgical procedure, or absence of functional loss. It is important to encourage all patients, particularly those with ocular hypertension, to seek continuous to monitor their condition.

The Affordable Care Act (ACA) was enacted in March 2010 with its primary goals being to make affordable health insurance available to more people and to generally lower the cost of health care [63]. Better health outcomes in Blacks have been linked to increases in health insurance coverage under the ACA [64]. While uninsured rates were reduced, Black Americans remained 1.5 times more likely to be uninsured than non-Hispanic white Americans [65]. Additionally, data gained from the National Health Interview Survey conducted between 2014–2016 revealed that access and utilization of eye care is lower among racial and ethnic minorities [66]. Increased access to health care and affordable insurance may improve the health outcomes of vulnerable populations with ocular hypertension.

6. Patient education for ocular hypertension

Patient education is an interactive process in which learning may take place between the healthcare provider and the patient. Increased patient education of vision health may lead to an increasing trend of eye doctor visits. Previous studies have reported that those with more education are more likely to seek care from an eye care professional as opposed to those with less education [67]. As a result of ocular hypertension, many Black patients were reported to present to an ophthalmologist with more extensive damage to the optic nerve as compared to whites [68]. As a result, the disease progression in Blacks was more vulnerable to malignancy even after intervention is initiated. Safer earlier cataract surgery and trabecular bypass are important treatment options that should be offered earlier.

Educating patients on ocular hypertension involves sharing the risk factors associated with the eye condition such as family history, age, medical conditions, and past eye injuries, as well prevention and treatment options. Due to the asymptomatic presentation of ocular hypertension with no signs of vision loss, it is possible that patients have not/will not seek treatment until further damage and vision loss occur. Prior recorded interactions between physicians and patients have found that providers were less likely to educate Black patients about glaucoma and were also less likely to educate patients of lower health literacy about glaucoma medications [69]. It is important for the patient's eye health that ocular hypertension and its potential progression to glaucoma are described and apprehensible, particularly to those in populations most at risk. Through patient education of ocular hypertension, the patient may better understand their susceptibility to eye disease and can seek early treatment if necessary.

7. Prevention and implementation of changes to address ocular hypertension in blacks

Given the information presented in this chapter, initiation of treatment for ocular hypertension may be started earlier in Blacks with the possibility of arresting or reducing elevated IOP. The aging population of adults aged 65 and older is continuously increasing with expectation of this number to reach nearly 90 million in the US by 2050 [70]. In addition, growing levels of obesity increasing the prevalence of diabetes make an increasing number of individuals at risk for vision loss in the future. As the risk factors for ocular hypertension increase, recognition of patient vulnerabilities and systemic level changes are needed to ensure that the needs of patients are properly and conveniently addressed.

8. Conclusion

This chapter has demonstrated the unique demographic and ocular characteristics that have affected Blacks in the progression of ocular hypertension. The combination of race, socioeconomic status, and access to treatment may influence the diagnosis and health outcome of individuals with ocular hypertension. Acknowledging these factors and implementing changes to promote early diagnosis and treatment, as well as addressing health and wealth disparities in high-risk populations, can lead to lower rates of glaucoma and blindness. Physician advice through patient education, as well as affordability, continuity, and frequent access to care has demonstrated a strong association with increased eye care services [71]. Diagnosis and early intervention of elevated levels of intraocular pressure and ocular hypertension may reduce the risk of glaucoma, vision loss, and blindness in future patients.

Conflict of interest

The authors declare no conflict of interest.

Author details

Daniel Laroche[1]* and Kara Rickford[2]

1 New York Eye and Ear Infirmary, Icahn School of Medicine of Mount Sinai, New York, NY, USA

2 School of Medicine, New York Medical College, Valhalla, NY, USA

*Address all correspondence to: dlarochemd@aol.com

References

[1] Sommer, A. Glaucoma: Facts and fancies. Eye 1996;10 295-301.

[2] Kass M, Heuer D, Higginbotham EJ, et al. The Ocular Hypertension Treatment Study: a randomized trail determines that topical ocular hypotensive medication delays or prevents the onset of primary open angle glaucoma. Archives of Ophthalmology. 2002;120 701-713.

[3] The Advanced Glaucoma Intervention Study (AGIS): 7. The relationship between control of intraocular pressure and visual field deterioration. American journal of ophthalmology. 2000;130 429-440.

[4] United States Census Bureau. Race and Ethnicity [Internet] 2019. Available from: https://data.census.gov/cedsci/profile?q=United%20States&g=0100000US [Accessed 2021-01-28]

[5] National Eye Institute. Eye Health Among African Americans. [Internet] 2019. Available from https://www.nei.nih.gov/learn-about-eye-health/resources-for-health-educators/eye-health-among-african-americans [Accessed 2021-01-28]

[6] The World Bank. Countries and Economies Data [Internet] 2020. Available from https://data.worldbank.org/country [Accessed 2021-02-07]

[7] World Health Organization. Eye Health [Internet] 2020. Available from https://www.afro.who.int/health-topics/eye-health#:~:text=Approximately%2026.3%20million%20people%20in,blind%20population%20reside%20in%20Africa. [Accessed 2021-02-07]

[8] Leske M, Connell A, Schachat A, Hyman L. The Barbados Eye Study. Prevalence of open angle glaucoma. Archives of Ophthalmology. 1994;112(6):821-829. DOI: 10.1001/archopht.1994.01090180121046

[9] Tielsch JM, Katz J, Singh K, et al. A population-based evaluation of glaucoma screening: the Baltimore Eye Survey. American Journal of Epidemiology. 1991;134(10):1102-1110. DOI: 10.1093/oxfordjournals.aje.a116013

[10] Wilson M, Gallardo M. Glaucoma Risk Factors: Ethnicity and Glaucoma. In: Schacknow P, Samples J, editors. The Glaucoma Book. 1st ed. New York: Springer; 2010. p. 101-107.

[11] Friedman D, Wilson M, Liebmanm J, et al. An evidence-based assessment of risk factors for the progression of ocular hypertension and glaucoma. American Journal of Ophthalmology. 2004;138(3):19-31. DOI: 10.1016/j.ajo.2004.04.058

[12] van Koolwijk LM, Ramdas WD, Ikram MK, et al. Common genetic determinants of intraocular pressure and primary open-angle glaucoma. PLoS Genetics. 2012;8(5):e1002611. DOI: 10.1371/journal.pgen.1002611

[13] McKean-Cowdin R, Fairbrother-Crisp A, Torres M, Lastra C, et al.The African American Eye Disease Study: Design and Methods. Ophthalmic Epidemiology. 2018;25(4):306-314. DOI: 10.1080/09286586.2018.1454965

[14] Laroche D, Nkrumah G, Ng C. Real-world efficacy of the Hydrus microstent in Black and Afro-Latinx patients with glaucoma: a retrospective study. Therapeutic Advances in Ophthalmology. 2020;12:2515841420964311. DOI: 10.1177/2515841420964311.

[15] Lahola-Chomiak AA, Walter MA. Molecular Genetics of Pigment Dispersion Syndrome

and Pigmentary Glaucoma: New Insights into Mechanisms. Journal of Ophthalmology. 2018;2018:5926906. DOI: 10.1155/2018/5926906.

[16] Jiao X, Yang Z, Yang X, et al. Common variants on chromosome 2 and risk of primary open-angle glaucoma in the Afro-Caribbean population of Barbados. Proceedings. 2009;106(40):17105-10. DOI: 10.1073/pnas.0907564106

[17] Chan C, Wiggs J, Sobrin L. The Genetic Influence on Corticosteroid-Induced Ocular Hypertension: A Field Positioned for Discovery. American Journal of Ophthalmology. 2019;202:1-5. DOI: 10.1016/j.ajo.2019.02.001

[18] Choquet H, Paylakhi S, Kneeland S, et al. A multiethnic genome-wide association study of primary open-angle glaucoma identifies novel risk loci. Nature Communications 2018;9:2278. DOI: 10.1038/s41467-018-04555-4

[19] Burdon KP, Macgregor S, Hewitt AW, et al. Genome-wide association study identifies susceptibility loci for open angle glaucoma at TMCO1 and CDKN2B-AS1. Nature Genetics. 2011;43(6):574-8. DOI: 10.1038/ng.824.

[20] Vithana EN, Khor CC, Qiao C, et al. Genome-wide association analyses identify three new susceptibility loci for primary angle closure glaucoma. Nature Genetics. 2012t;44(10):1142-1146. DOI: 10.1038/ng.2390.

[21] Masakazu N, Yoko I, Takazumi T, et al. Three susceptible loci associated with primary open-angle glaucoma identified by genome-wide association study in a Japanese population. 2009;106(31):12838-12842. DOI: 10.1073/pnas.0906397106

[22] Khachatryan N, Pistilli M, Maguire MG, et al. Primary Open-Angle African American Glaucoma Genetics (POAAGG) Study: gender and risk of POAG in African Americans. PLoS One. 2019;14(8):e0218804. DOI: 10.1371/journal.pone.0218804.

[23] Gudiseva H, Verma S, Chavali V, et al. Genome wide-association study identifies novel loci in the Primary Open-Angle African American Glaucoma Genetics (POAAGG) study. DOI 10.1101/2020.02.27.968156

[24] Vranka JA, Kelley MJ, Acott TS, Keller KE. Extracellular matrix in the trabecular meshwork: intraocular pressure regulation and dysregulation in glaucoma. Experimental Eye Research. 2015;133:112-25. DOI: 10.1016/j.exer.2014.07.014.

[25] Ventura A, Bohnke M, Mojon D. Central corneal thickness measurements in patients with normal tension glaucoma, primary open angle glaucoma, pseudoexfoliation glaucoma, or ocular hypertension. British Journal of Ophthalmology. 2001;85:792-795.

[26] Medeiros FA, Weinreb RN. Is corneal thickness an independent risk factor for glaucoma? Ophthalmology. 2012;119(3):435-6. DOI: 10.1016/j.ophtha.2012.01.018.

[27] Miglior S, Zeyen T, Pfeiffer N, et al. European Glaucoma Prevention Study (EGPS) Group. Results of the European Glaucoma Prevention Study. Ophthalmology. 2005;112(3):366-75. DOI: 10.1016/j.ophtha.2004.11.030.

[28] Gordon M, Torri V, Miglior S, et al. Validated prediction model for the development of primary open-angle glaucoma in individuals with ocular hypertension. Ophthalmology. 2007;114(1):10-9. DOI: 10.1016/j.ophtha.2006.08.031.

[29] Copt R, Thomas R, Memoud A. Corneal Thickness in Ocular Hypertension, Primary Open-angle Glaucoma, and Normal Tension

Glaucoma. Archives of Ophthalmology. 1999;117(1):14-16. DOI: 10.1001/archopht.117.1.14

[30] Herman D, Hodge D, Bourne W. Increased Corneal Thickness in Patients With Ocular Hypertension. Archives of Ophthalmology. 2001;119(3):334-336. DOI:10.1001/archopht.119.3.334

[31] Baboolal SO, Smit DP. South African Eye Study (SAES): ethnic differences in central corneal thickness and intraocular pressure. Eye (Lond). 2018;32(4):749-756. DOI: 10.1038/eye.2017.291.

[32] Girkin CA. Differences in optic nerve structure between individuals of predominantly African and European ancestry: Implications for disease detection and pathogenesis. Clinical Ophthalmology. 2008;2(1):65-9. doi: 10.2147/opth.s1628.

[33] Varma R, Tielsch JM, Quigley HA, Hilton SC, Katz J, Spaeth GL, Sommer A. Race-, age-, gender-, and refractive error-related differences in the normal optic disc. Archives of Ophthalmology. 1994;112(8):1068-76. doi: 10.1001/archopht.1994.01090200074026.

[34] Chi T, Ritch R, Stickler D, Pitman B, Tsai C, Hsieh FY. Racial differences in optic nerve head parameters. Archives of Ophthalmology. 1989;107(6):836-9. DOI: 10.1001/archopht.1989.01070010858029.

[35] Fortune B, Cull G, Reynaud J, Wang L, Burgoyne CF. Relating Retinal Ganglion Cell Function and Retinal Nerve Fiber Layer (RNFL) Retardance to Progressive Loss of RNFL Thickness and Optic Nerve Axons in Experimental Glaucoma. Investigative Ophthalmology & Visual Science. 2015;56(6):3936-44. doi: 10.1167/iovs.15-16548.

[36] Bowd C, Weinreb R, Williams J, et al. The Retinal Nerve Fiber Layer Thickness in Ocular Hypertensive, Normal, and Glaucomatous Eyes With Optical Coherence Tomography. Archives of Ophthalmology. 2000;118(1):22-26. doi:10.1001/archopht.118.1.22

[37] Zangwill L, van Horn S, De Souza M, et al. Optic Nerve Head Topography in Ocular Hypertensive Eyes Using Confocal Scanning Laser Ophthalmoscopy. American Journal of Ophthalmology. 1996;122(4):520-525.

[38] The International Agency for the Prevention of Blindness. Vision 2020 [Internet] 2020. Available from: https://www.iapb.org/about/vision-2020/ [Accessed 02-01-2021]

[39] Azuara-Blanco A, Burr J, Ramsay C, et al. Effectiveness of early lens extraction for the treatment of primary angle-closure glaucoma (EAGLE): a randomised controlled trial. Lancet. 2016;388(10052):1389-1397. DOI: 10.1016/S0140-6736(16)30956-4.

[40] Bazargan M, Ekwegh T, Cobb S, Adinkrah E, Assari S. Eye Examination Recency among African American Older Adults with Chronic Medical Conditions. Healthcare (Basel). 2020;8(2):94. DOI: 10.3390/healthcare8020094.

[41] The Century Foundation. Racism, Inequality, and Health Care for African Americans [Internet] 2019. Available from: https://tcf.org/content/report/racism-inequality-health-care-african-americans/?session=1&session=1 [Accessed 02-01-2021]

[42] Brookings. Examining the Black-white wealth gap [Internet] 2020. Available from: https://www.brookings.edu/blog/up-front/2020/02/27/examining-the-black-white-wealth-gap/ [Accessed 02-07-2021]

[43] American Academy of Ophthalmology. Eye Health Statistics

[Internet] 2021. Available from: https://www.aao.org/newsroom/eye-health-statistics [Accessed 02-01-2021]

[44] Sarrazin MS, Campbell ME, Richardson KK, Rosenthal GE. Racial segregation and disparities in health care delivery: conceptual model and empirical assessment. Health Services Research. 2009;44(4):1424-44. DOI: 10.1111/j.1475-6773.2009.00977

[45] Scharff DP, Mathews KJ, Jackson P, Hoffsuemmer J, Martin E, Edwards D. More than Tuskegee: understanding mistrust about research participation. Journal of Health Care for the Poor and Underserved. 2010;21(3):879-97. DOI: 10.1353/hpu.0.0323.

[46] Nuriddin A, Mooney G, White AIR. Reckoning with histories of medical racism and violence in the USA. Lancet. 2020;396(10256):949-951. DOI: 10.1016/S0140-6736(20)32032-8.

[47] American Bar Association. The Impact of Structural Racism in Employment and Wages on Minority Women's Health [Internet] 2018. Available from: https://www.americanbar.org/groups/crsj/publications/human_rights_magazine_home/the-state-of-healthcare-in-the-united-states/minority-womens-health/ [Accessed 02-01-2021]

[48] The Federal Reserve System. Disparities in Wealth by Race and Ethnicity in the 2019 Survey of Consumer Finances [Internet] 2020. Available from: https://www.federalreserve.gov/econres/notes/feds-notes/disparities-in-wealth-by-race-and-ethnicity-in-the-2019-survey-of-consumer-finances-20200928.htm [Accessed 02-01-2021]

[49] Health Resources & Services Administration. MUA Find [Internet] 2021. Available from: https://data.hrsa.gov/tools/shortage-area/mua-find#:~:text=Medically%20Underserved%20Areas%2FPopulations%20are,or%20a%20high%20elderly%20population. [Accessed 02-01-2021]

[50] Ending systemic racism in medicine. Nature Medicine. 2020;26:985. DOI: 10.1038/s41591-020-0993-2

[51] Kuczmarski AV, Cotugna N, Mason MA, Evans MK, Zonderman AB. Depression and Cognitive Impairment Are Associated with Low Education and Literacy Status and Smoking but Not Caffeine Consumption in Urban African Americans and White Adults. Journal of caffeine research. 2015;5(1):31-41. DOI: 10.1089/jcr.2014.0019

[52] Tseng YC, Liu SH, Lou MF, Huang GS. Quality of life in older adults with sensory impairments: a systematic review. Quality of Life Research. 2018;27(8):1957-1971. DOI: 10.1007/s11136-018-1799-2.

[53] Swenor BK, Lee MJ, Varadaraj V, Whitson HE, Ramulu PY. Aging With Vision Loss: A Framework for Assessing the Impact of Visual Impairment on Older Adults. Gerontologist. 2020;60(6):989-995. DOI: 10.1093/geront/gnz117.

[54] Vu HT, Keeffe JE, McCarty CA, Taylor HR. Impact of unilateral and bilateral vision loss on quality of life. British Journal of Ophthalmology. 2005;89(3):360-3. DOI: 10.1136/bjo.2004.047498.

[55] National Academies of Sciences, Engineering, and Medicine. Health-Care Utilization as a Proxy in Disability Determination [Internet] 2018. Available from: https://www.ncbi.nlm.nih.gov/books/NBK500102/ doi: 10.17226/24969 [Accessed 02-01-2021]

[56] Sommer A, Tielsch JM, Katz J, Quigley HA, Gottsch JD, Javitt JC, Martone JF, Royall RM, Witt KA, Ezrine S. Racial differences in the

cause-specific prevalence of blindness in east Baltimore. New England Journal of Medicine. 1991;325(20):1412-7. DOI: 10.1056/NEJM199111143252004.

[57] Johnson CA, Keltner JL, Cello KE, et al. Baseline visual field characteristics in the ocular hypertension treatment study. Ophthalmology. 2002;109(3):432-7. DOI: 10.1016/s0161-6420(01)00948-4

[58] Laroche D, Nkrumah G, Ng C. Real-world efficacy of the Hydrus microstent in Black and Afro-Latinx patients with glaucoma: a retrospective study. Therapeutic Advances in Ophthalmology 2020;12 1-9.

[59] Heijl A, Leske MC, Bengtsson B, et al. Reduction of intraocular pressure and glaucoma progression: results from the Early Manifest Glaucoma Trial. Archives of Ophthalmology. 2002;120(10):1268-79. doi: 10.1001/archopht.120.10.1268.

[60] Washington DL. Charting the path from lack of insurance to poor health outcomes. Western Journal of Medicine. 2001;175(1):23. DOI: 10.1136/ewjm.175.1.23.

[61] Institute of Medicine (US) Committee on the Consequences of Uninsurance. Care Without Coverage: Too Little, Too Late. Effects of Health Insurance on Health [Internet] 2002. Available from: https://www.ncbi.nlm.nih.gov/books/NBK220636/ [Accessed 02-01-2021]

[62] Muñoz B, West SK, Rubin GS, et al. Causes of blindness and visual impairment in a population of older Americans: The Salisbury Eye Evaluation Study. Archives of Ophthalmology. 2000;118(6):819-25. DOI: 10.1001/archopht.118.6.819.

[63] HealthCare.Gov. Affordable Care Act (ACA) [Internet] 2020. Available from: https://www.healthcare.gov/glossary/affordable-care-act [Accessed 02-01-2021]

[64] Sommers BD, McMurtry CL, Blendon RJ, et al. Beyond Health Insurance: Remaining Disparities in US Health Care in the Post-ACA Era. Milbank Quarterly. 2017;95(1):43-69. DOI: 10.1111/1468-0009.12245.

[65] Kaiser Family Foundation. Changes in Health Coverage by Race and Ethnicity since the ACA, 2010-2018 [Internet] 2020. Available from: https://www.kff.org/racial-equity-and-health-policy/issue-brief/changes-in-health-coverage-by-race-and-ethnicity-since-the-aca-2010-2018/ [Accessed 02-01-2021]

[66] Varadaraj V, Frick KD, Saaddine JB, et al. Trends in Eye Care Use and Eyeglasses Affordability: The US National Health Interview Survey, 2008-2016. JAMA Ophthalmology. 2019;137(4):391-398. DOI: 10.1001/jamaophthalmol.2018.6799.

[67] Orr P, Barrón Y, Schein OD, et al. Eye care utilization by older Americans: the SEE Project. Salisbury Eye Evaluation. Ophthalmology. 1999;106(5):904-9. DOI: 10.1016/s0161-6420(99)00508-4.

[68] Greenidge KC, Dweck M. Glaucoma in the black population: a problem of blindness. Journal of the National Medical Association. 1988;80(12):1305-9.

[69] Sleath B, Blalock SJ, Carpenter DM, et al. Provider Education about Glaucoma and Glaucoma Medications during Videotaped Medical Visits. Journal of Ophthalmology. 2014;2014:238939. DOI: 10.1155/2014/238939.

[70] Population Reference Bureau. America's Aging Population [Internet] 2011. Available from: https://assets.

prb.org/pdf11/aging-in-america.pdf
[Accessed 02-01-2021]

[71] Unzueta M, Globe D, Wu J, et al. Compliance with recommendations for follow-up care in Latinos: the Los Angeles Latino Eye Study. Ethnicity & Disease. 2004;14(2):285-91.

Chapter 2

Retinal Vascular Implications of Ocular Hypertension

Fidan Jmor and John C. Chen

Abstract

In this chapter, we review the basics of retinal vascular anatomy and discuss the physiologic process of retinal blood flow regulation. We then aim to explore the relationship between intraocular pressure and retinal circulation, taking into account factors that affect retinal hemodynamics. Specifically, we discuss the concepts of ocular perfusion pressure, baro-damage to the endothelium and transmural pressure in relation to the intraocular pressure. Finally, we demonstrate the inter-relationships of these factors and concepts in the pathogenesis of some retinal vascular conditions; more particularly, through examples of two common clinical pathologies of diabetic retinopathy and central retinal vein occlusion.

Keywords: retinal hemodynamics, baro-damage, ocular perfusion pressure, diabetic retinopathy, central retinal vein occlusion, blood flow

1. Introduction

The retina shares similar anatomical features and physiological properties with other end organs such as the brain and the kidney, namely the presence of blood–brain, blood-kidney and blood-retina barrier as well as non-anastomotic end arteries [1]. Retinal fundoscopy and digital imaging have allowed for retinal microvascular abnormalities to be directly and non-invasively identified and studied as a means of better understanding the manifestation of systemic microcirculatory disorders.

Although not yet completely understood, hemodynamic factors such as perfusion pressure, blood viscosity, vascular resistance and the variations on vessel caliber that ensue, determine the blood supply and flow to the retina. By understanding these processes and their disturbances, we can better characterize the pathological processes that occur in many ocular and systemic diseases such as glaucoma, age-related macular degeneration and diabetic retinopathy, to name a few.

Retinal hemodynamics are influenced by a number of factors. Blood flow, arterial and venous pressure, vascular resistance, and blood viscosity all play important roles. Our understanding of their inter-relationship is derived from concepts used in fluid flow systems borrowed from engineering and from other physiologic studies of blood flow. Mathematically, this relationship is often simplified into Poisseuille's equation, which we will discuss more in detail in the diabetic retinopathy section. However, it is important to know that Poisseuille's equation is used mainly for Newtonian fluid in a system with laminar flow. In the retinal arterioles, the flow is often turbulent; and blood itself is not truly a Newtonian fluid. In this chapter, we will not delve into the detailed discussions of mathematical modeling, but will only use them to better understand the hemodynamics and its biologic consequences.

Clinical evidence for the relationship between intraocular pressure (IOP) and retinal hemodynamics remains inconsistent and difficult to interpret. The reasons are two-fold. Firstly, it is difficult to capture the data of multiple variables that may contribute to hemodynamics in a clinical setting. Second, in a complex system such as retinal circulation, many variables act as both dependent and as independent variables in many feed-back loops. Moreover, these feed-back loops may vary depending on the underlying disease processes in individuals with multiple co-morbidities [2, 3].

2. Blood supply to the posterior segment

2.1 Retinal blood supply

The metabolic demands and the oxygen requirement of the retina are met by two distinct vascular systems. The inner two thirds of the retina is supplied by inherent intra-retinal vessels, fed by the central retinal artery, and drained via the central retinal vein. The photoreceptors and outer one third of the retina are supplied by the choroidal circulation [4]. Both of these supplies originate from the ophthalmic artery, which itself is a branch of the internal carotid artery.

The central retinal artery traverses through the orbital portion of the optic nerve, entering the optic disc through the lamina cribrosa. At the optic nerve, there is a combination of choroidal and retinal arterial circulation, details of which are dealt with in other chapters. The central retinal artery branches into four principal intra-retinal arteries. These further bifurcate into increasingly smaller arterioles, feeding eventually into a capillary bed as they extend towards the peripheral retina (**Figure 1**).

These capillary beds form interconnecting networks linking terminal branches of pre-capillary arterioles and post-capillary venules. Although in the juxta-papillary region this is arranged in three layers, the peri-macular region has two layers; a superficial layer located in the nerve fiber and ganglion cell layers and a second, deeper layer in the inner nuclear and outer plexiform layers. With the exception of a small avascular rim, both superficial and deep plexi reach almost to the edge of the human retina [6, 7]. The fovea is also avascular, receiving adequate oxygenation via the choroidal circulation [8].

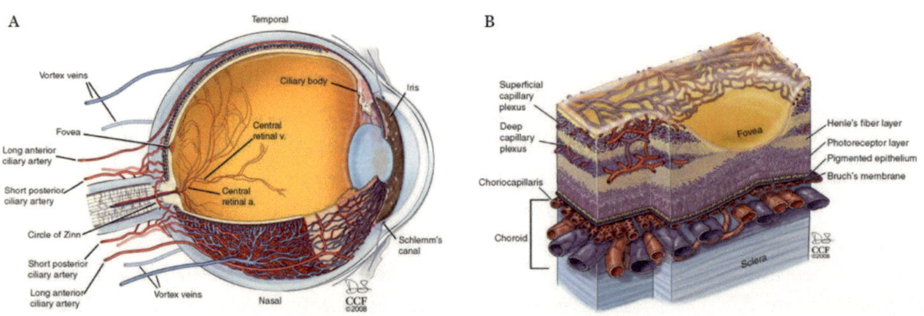

Figure 1.
(A) Sagittal drawing of the human eye showing the retinal and choroidal circulation of the left eye. (B) Cross-sectional drawing of the retinal and choroidal vasculature at the level of the fovea. Adapted from Anand-Apte and Hollyfield with drawings by Dave Schumick [5].

Post-capillary venules feed back into the superior and inferior hemi-central retinal veins, eventually uniting into the central retinal vein which centralizes at variable depths within the optic nerve eventually draining into the cavernous sinus.

The neighboring endothelial cells lining the retinal vasculature form tight junctions. Together with pericytes, they form a highly selective semi-permeable border that prevents solutes in the circulating blood from non-selectively crossing into the interstitial space within the neuroretina, constituting the inner blood retina barrier.

2.2 Choroidal blood supply

The choroid has approximately 80% of the total ocular blood supply relative to iris-ciliary body and retina [9] and consists of three distinct layers of gradually decreasing vessel caliber; Haller's layer comprises the outer, larger sized vessels, Sattler's layer is intermedial with medium-sized vessels, and the deeper choriocapillaris contains vessels with the smallest diameter [10]. The anterior choroid is supplied by the long ciliary arteries, whereas the posterior choroid is supplied by the short posterior ciliary arteries. The entire choroid drains into the vortex veins [11].

Unlike the retinal vasculature, choroidal capillaries are fenestrated, allowing free passage and exchange of intravascular contents and interstitial space, including macromolecules and cellular components. It is the monolayer of retinal pigment epithelial cells, with tight junctions at the apical aspect, that form the outer blood-retinal barrier (**Figure 1**).

2.3 Blood supply to the optic nerve head

The blood supply to the optic nerve head is complex, deriving from both the central retinal artery and from the choroid through the short posterior ciliary arteries. It has been specifically discussed in an earlier chapter in this book.

3. Blood flow within the posterior segment

3.1 Retinal blood flow

The human retina is a metabolically demanding tissue. Tissue damage and cell death can be brought about by small alterations in oxygenation or blood flow; hypoperfusion leads to hypoxia and ischemic damage; [12] whilst hyperperfusion and/or high oxygen tension leads to formation of reactive oxygen species, leading to oxidative damage [13]. Retinal blood flow must necessarily be highly regulated, and is dependent on the relationship between perfusion pressure and local resistance [11].

Under physiological conditions, retinal arterial pressure is more or less equal to mean arterial blood pressure and retinal venous pressure is more or less equal to the IOP. The difference between these pressures constitutes the driving force propelling blood through ocular capillary beds. In general, mean ocular perfusion pressure (MOPP) is positively correlated with arterial blood pressure and negatively correlated with IOP [14].

MOPP, defined as the difference between two-thirds of the mean arterial pressure (MAP) and the IOP, is a clinically modifiable factor in diseases such as Diabetic retinopathy (DR) where there is altered tissue perfusion. This will be reviewed in more detail later in the chapter.

The normal retinal hemodynamic response to increases in perfusion pressure is an increase in vascular resistance, [15] otherwise referred to as the myogenic response. This behavior is intrinsic to smooth muscle cells such as those that line retinal arterioles and is independent of metabolic and hormonal influences.

3.2 Choroidal blood flow

Whilst retinal blood flow is characterized by a low perfusion rate, a high vascular resistance and a high oxygen extraction, the choroid, by contrast, shows a low vascular resistance, high perfusion rate and low oxygen extraction [11]

4. Regulation of blood flow in the posterior segment

4.1 Retinal blood flow regulation

Ordinarily, blood flow is regulated in response to changes in perfusion pressure and tissue oxygen tension.

Retinal circulation differs from blood flow in other non-neural systems in that local neural activity can evoke localized changes in blood flow. This behavior, termed neurovascular coupling, has been observed in the retina [16, 17] and is an emerging area of research in glaucoma [18].

More importantly, the regulatory effect of perfusion pressure on blood flow is blunted through the process of autoregulation. That is, the retinal blood flow, like that of the blood flow in the brain, is maintained or stabilized through a wide range of variations of perfusion pressure. Previous studies have shown that retinal autoregulation is adequately compensated in experimental elevations of IOP up to 29 mmHg, [19] whilst the retinal vasculature behaves more passively for greater increases in IOP [20]. This autoregulatory response is also noted in incidences of increased perfusion pressure such as periods of dynamic and static exercise; where a rise in perfusion pressure of up to 34% rise results in a rise in flow of only 4–8% [21, 22]. Teleologically, autoregulation is a protective mechanism to maintain a steady blood flow to the retina

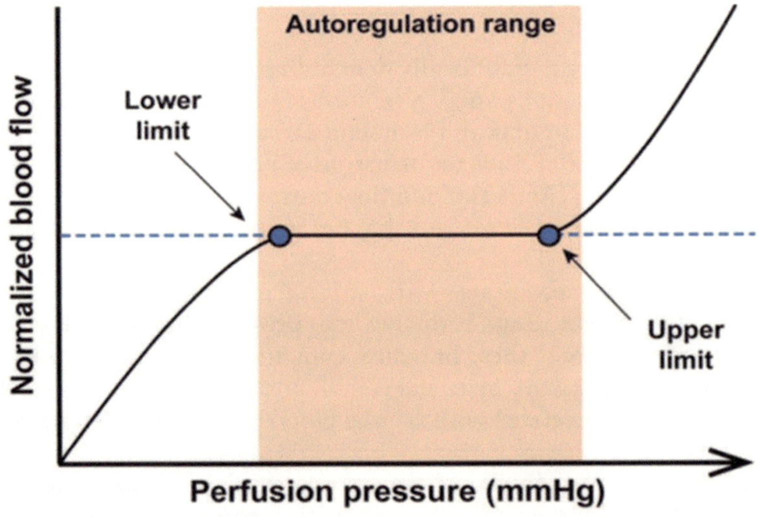

Figure 2.
Schematic of blood flow autoregulation in the eye. When fluctuations in ocular prefusion pressure exceed the autoregulation range defined by this plateau, vasomotor adjustments are incomplete and blood flow changes passively as ocular perfusion pressure changes. Figure adapted with permission from Wareham and Calkins [18].

to satisfy its metabolic demands which changes very little as compared to the wide swings of systemic blood pressure and even diurnal variations of IOP.

The autoregulation curve (**Figure 2**) shows how blood flow changes in response to ocular perfusion pressure. The curve includes a plateau region across a range of ocular perfusion pressures where the blood flow is fully compensated by the above mentioned autoregulatory mechanisms.

4.2 Choroidal blood flow regulation

Choroidal blood flow regulation is distinctly different from that of the retina; [23]. Firstly, the choroidal vascular bed is extensively innervated [24] though only partly autoregulated. There is little to no oncotic pressure gradient between intra- and extravascular spaces within the choroid due to capillary fenestration. This lack of oncotic pressure coupled with the absence of lymphatic vessels, allows for the onset of choroidal effusion if the IOP drops below a certain level.

5. Effect of intraocular pressure on retinal hemodynamics

The high metabolic and oxygen demand posed by retinal tissue is met by maintaining a steady blood flow. IOP is a major determinant of both retinal vascular perfusion pressure as well as vascular transmural pressure.

Over the past two decades, elegant mathematical modeling has been used to assess the effect of IOP elevation on the lamina cribrosa [25–27] and on arteriovenous distribution within the retinal microvasculature [28, 29].

Clinical evidence of the impact of IOP on retinal blood flow and velocity is inconsistent. Whilst several clinical studies have shown that as IOP increases, retinal and retrobulbar blood flow decreases, [30–32] others have not found this to hold true in various settings including post-operative trabeculectomy patients [33] and in patients treated with IOP lowering medications [34–36]. These inconsistencies are likely due to numerous factors, including arterial blood pressure and blood flow autoregulation [2] and the intrinsic difficulty of evaluating the individual contribution of these factors in a clinical setting. In organs where autoregulation is maintained, perfusion pressure is a weak parameter in altering blood flow within that organ.

In disease conditions, when autoregulation is disturbed, or absent, perfusion pressure as a determinant of blood flow becomes paramount. The role of arterial pressure and IOP can be easily understood in the following ways;

1. When the perfusion pressure is too high, there will be increased blood flow as well as increased transmural pressure within the retinal vasculature. This leads to endothelial damage (baro-damage) and the breakdown of the blood-retinal barrier. The endothelial damage, together with increased transmural pressure, leads to exudation and extravasation of blood products, causing retinal edema. Clinically, this series of events can be caused by an acute rise in blood pressure such as seen in malignant hypertension. It can also be seen after ocular filtration surgery when there is a significant reduction in IOP outside the range of autoregulation. In hypotony maculopathy, for example, we may often note optic disc swelling and tortuous vessels, and also macular edema and subretinal fluid, albeit more rarely.

2. Conversely, when the perfusion pressure is too low, the blood flow will fall below that required to maintain retinal metabolic and oxygen demand, thus leading to ischemia. This is seen in a number of examples:

i. In acute central retinal artery occlusion (CRAO) when an embolus blocks the retinal circulation of the central retinal artery at the optic disc thus leading to reduced blood flow and therefore widespread retinal ischemia. Anecdotal evidence from clinical experience has shown that reducing the IOP may aid in reversing the occlusion, such as when performing an anterior chamber paracentesis to rapidly reduce the IOP [37].

ii. In some patients with aortic stenosis or with carotid artery occlusion, perfusion pressure at the optic nerve head can become significant lower and fall outside the autoregulatory range. Any variations in systemic blood pressure or IOP may then cause a large enough drop in blood flow within the retinal vasculature, leading to ischemic damage to the retina ranging from mild form such as paracentral acute middle maculopathy (PAMM) or severe ischemia such as CRAO.

Furthermore, the picture may be complicated in cases where a state of low perfusion is maintained at the limits of autoregulation on a more chronic basis. Here there is no sudden event causing an ischemic infarct such as is seen in CRAO or PAMM, but rather there may be enough persistent venous stasis to induce a clinical picture reminiscent of Central retinal vein occlusion (CRVO), in a condition commonly known as Ocular ischemic syndrome or venous stasis retinopathy.

6. Effect of ocular hypertension in clinical retinopathies

The effect of IOP in retinal vascular diseases is well documented. Ocular hypertension (OHT) is a risk factor for the development of CRVO. On the other hand, OHT seems to have a protective effect on the development and progression of DR.

6.1 Retinal vein occlusions

Retinal vein occlusion (RVO) is one of the most common retinal vascular disorders, with a prevalence as high as 4.6% in adults 80 years or older, [38] and can be seen in the central retinal vein (CRVO) or in branched veins (BRVO). Although the underlying mechanisms governing RVOs are multifactorial, Virchow's triad teaches that a combination of blood flow stasis, endothelial cell damage and hypercoagulability leads to thrombosis [39].

In CRVOs, the proximity of the central retinal vein and artery (enveloped in a common fibrous tissue) to one another within the optic disc means that the presence of arterial disease such as systemic hypertension and arteriosclerosis, can predispose the central retinal vein to a pre-morbid low flow state. Additional risk factors for RVOs include coagulation disorders and hyper viscosity states, but one of the most frequently encountered risk factors is glaucoma/OHT. In BRVOs, hemodynamic changes occur at arteriovenous crossings coupled with altered blood flow thus leading to localized venous compression [40].

Verhoeff first described the relationship between glaucoma and CRVO in 1913, where he postulated that increased IOP compresses and collapses the wall of the central retinal vein, leading to intimal proliferation within the vein [41]. Whereas primary open angle glaucoma (POAG) is a feature which precedes CRVO in between 10 and 40% of patients, [Larsson] the prevalence of POAG in BRVO appears to be less frequent at between 6 and 15% [42, 43].

Many studies have analyzed the relationship between glaucoma and RVO risk, with contradictory outcomes. Yin et al. recently performed a meta-analysis of research conducted between 1977 and 2015, examining the relationship between glaucoma and RVO and found glaucoma to be a core risk factor for RVO in 15 studies with high methodological quality [44]. The studies reviewed have suggested a number of different potential hypotheses;

- Some have considered that glaucoma and RVO are both manifestations of an underlying vascular abnormality or vascular dysregulation. Both conditions share similar risk factors; hyperlipidemia, smoking habits, abnormal plasma viscosity and inflammatory activity [45–47]. Other abnormalities found with raised fibrinogen and IgA are also consistent, implying that associated medical conditions are of greater significance to deterioration in RVOs rather than the presence of glaucoma or OHT itself.

- POAG has been found to precede vascular occlusion [48–51] and disc hemorrhage is frequently seen in POAG patients in the clinic, leading to the hypothesis that RVO and POAG might share a common pathogenesis [52].

- In patients with compromised retinal circulation and autoregulatory mechanisms, a minor rise in IOP may be high enough to reduce ocular blood flow sufficiently to propagate stagnant blood flow. This hypothesis is supported by the finding of a reduction in IOP thresholds, normally estimated to be 27-30 mmHg in patients with chronic simple glaucoma, up to which autoregulation, by increasing arterial pressure with decreased vascular resistance, maintains normal blood flow [53].

- Some groups have presented a vascular hypothesis of glaucoma; [54] individuals with primary angle closure glaucoma as well as POAG are known to have narrower retinal arteries and veins than normal subjects [55]. Therefore elevated IOP may lead to greater vessel wall compression in these "primed" vessels, leading to vein intimal proliferation and collapse of retinal capillaries [56, 57].

Although glaucoma and OHT are risk factors for CRVO, once CRVO is established, there is a curious phenomenon of lowered IOP in the eye with the RVO as compared to the fellow, non-affected eye. The exact cause of this lowering of IOP is not understood, though Hayreh et al. postulated it may be due to the release of soluble factors induced by relative ocular ischemia [58].

This lowered IOP, combined with increased venous pressure, increases the capillary hydrostatic pressure according to Starling's Law [59]. This in turn leads to increased leakage from compromised endothelial cells; the clinical presentation of which is seen as macular edema and hemorrhagic retinopathy.

In fluid dynamics, Bernoulli's principle states that an increase in the speed of a fluid (kinetic energy) occurs simultaneously with a decrease in static pressure or a decrease in the fluid's potential energy. The converse is the case in severe CRVO where the venous thrombosis causes significant slowing down of blood flow velocity, the kinetic energy that drives the blood flow forward within the capillary network can be reduced enough to cause intra capillary thrombosis, as evidenced in wide spread capillary drop out on fluorescein angiography. When enough retinal area is involved (usual criteria is 10 disc areas on angiography), the resulting ischemia may lead to neovascular complications. This is referred to clinically as ischemic CRVO. At the same time, according to Bernoulli principle of total energy conservation, the decrease in kinetic energy will result in a commensurate increase in the potential energy as expressed in lateral or transmural pressure. This increase in lateral pressure causes extravasation of fluid into the interstitial space and results in retinal edema.

Whilst our current approach to the management of RVO incorporates addressing associated systemic risks common to both glaucoma and RVO, further work is needed to establish the exact effects of elevated IOP as well as the additional effect of glaucoma medications on the autoregulatory capacity of retinal blood flow e.g. timolol enhancing autoregulation and thus possibly aiding perfusion following RVOs [60].

6.2 Diabetic retinopathy

DR continues to be one of the leading causes of blindness globally, and as retinal capillaries can be visualized directly, the progression of DR can be continuously assessed. The microangiopathic processes noted in the retina are echoed in the glomeruli of diabetic patients. The glomerular microcirculation has received significant attention as it's accessibility to both clinical and experimental observation is unique, in that fluid and macromolecule movement across capillary walls can be easily quantified [61]. Glomerular hemodynamic abnormalities, as with the microcirculation in the retina, are thought to be mediated by a complex chain of events including direct mechanical injury (baro-damage) to the capillaries, [62] and subsequent intracapillary coagulation [63]. As with renal microcirculation, hemodynamic abnormalities within the retinal microcirculation can be detected many years before DR becomes overt.

Multiple factors, including altered levels of vasoactive substances, altered vasomotor responsiveness and persistent hypoxia leads to marked venous vasodilation. Although the consequent elevations in capillary pressure and blood flow may be the inciting mechanism for the onset of diabetic microangiopathy, the factors linking hyperglycemia to vascular cell dysfunction, capillary dropout, tissue hypoxia, and abnormal angiogenesis, remain poorly described [64, 65]. Patients with diabetes are however known to have dysfunctional retinal perfusion [66] and an abnormal autoregulatory capacity [67].

The retinal microcirculation is sensitive to local variations in oxygen tension, with capillary blood flow and vessel diameter varying as necessary in response to local metabolic demand; [61] retinal capillaries dilate with low ambient oxygen tensions, and constrict with high ambient oxygen tensions [68].

As DR progresses, capillary microaneurysms, exudates and hemorrhages are seen, followed by endothelial proliferation with neovascularization. In the latter stages, focal retinal atrophy and vitreous body adhesions occur, eventually leading to tractional retinal detachment if not treated. These advancing stages of DR are characterized by local capillary basement membrane thickening, endothelial proliferation and intracapillary thrombosis, the latter of which further aggravates endothelial proliferation. This cascade results in capillary lumen obstruction, exacerbated even more so in the presence of hypoxia. Over time, retinal microvascular damage in the presence of persistent hypoxia results in elevated intraocular vascular endothelial growth factor (VEGF), an endothelial-specific diffusible factor that mediates permeability and development of vasculature.

That hemodynamic factors play an important role in the development of DR is evident in many clinical observations. Generally, disease states that cause an elevation in retinal perfusion pressure hastens the onset of DR. When retinopathy is already established, the same high perfusion state can cause a more rapid progression of the retinopathy. Conditions that cause perfusion pressure to increase include:

i. Systemic hypertension

The United Kingdom Prospective Diabetes study (UKPDS) found that improved blood pressure control decreased the progression of diabetic microangiopathy and correlated with a reduction in risk of cerebrovascular incidents by more than a third [69].

ii. Pregnancy

It is widely recognized that pregnancy worsens during pregnancy [70–72]. DR severity pre-pregnancy, metabolic control during pregnancy and pregnancy related hypertension have all be identified as risk factors for this worsening [70, 72]. Increased cardiac output and plasma volume during pregnancy, as well as a decrease in peripheral vascular resistance significantly increase blood flow to different parts of the body, including the retinal vasculature. Chen et al. have shown that when increased blood flow was documented in the first trimester, DR progressed, in contrast to unchanging DR severity in women whose retinal blood flow remained unchanged [73]. The hyperdynamic circulatory state induced by pregnancy is counteracted effectively only if normal autoregulatory control of blood flow is maintained. In some diabetic pregnant women, these mechanisms are flawed, where increased blood flow potentially inflicts endothelial damage by inciting additional shear stress at the capillary level.

Conversely, disease state that leads to a decrease in retinal perfusion pressure may protect against the development or the progression of DR.

6.2.1 Carotid stenosis

It is noted that patients with carotid artery disease may have eyes with asymmetric severity of DR; with the less affected eye having an ipsilateral carotid artery that is more obstructed. The protective effect has been attributed to the reduction in the retinal arterial perfusion pressure [74]. It should be noted that when the carotid occlusion becomes severe, that is exceeding 90% of the vessel caliber, this protective effect is lost due to consequent ischemia [75].

6.2.2 Optic atrophy

A similar asymmetric DR severity can be seen in patients with optic atrophy. The side with optic atrophy has severe DR compared to the side with a normal optic nerve. This can be explained by the narrowing of the retinal arterioles generally seen in eyes following the onset of optic atrophy [76, 77]. In these eyes, although the arterial pressure at the optic nerve head may not be altered, the capillary perfusion pressure is much reduced due to the narrowing of the vessel caliber.

The relation between vessel caliber and end capillary perfusion pressure can be understood through Poisseuille's blood flow equation: [78].

$$\Delta p = 8\mu LQ/\pi R4 = 8\mu LQ/A^2 \quad (1)$$

where:

Δp is the pressure difference between the two ends (pressure drop) L is the length of pipe (distance blood travels to reach capillary bed) μ is the dynamic viscosity.

Q is the volumetric flow rate (blood flow).

R is the pipe radius (vessel caliber).

A is the cross section of pipe (vessel cross sectional area).

Note that the drop in perfusion pressure at the capillary level is inversely correlated to the square of the cross sectional area of the vasculature. The narrower the vessel caliber, the smaller the cross sectional area, the drop in perfusion at the capillary level is increased by the fourth power of the radius of the vessel.

6.2.3 Myopia

In patients with anisometropia where one eye is significantly more myopic than the other, it is the eye with higher myopia that demonstrates less severe DR levels [79, 80]. Myopic eyes tend to have longer axial lengths, consequently for every corresponding point in the retina, the arteriole has to travel further to reach compared to that in an eye with a normal axial length. This increased blood travel, can be simplified as having comparatively longer vessel or pipe length, with a consequent decrease in end capillary perfusion. Mathematically, we can once again understand it through Poisseuille's equation: as L increases, Δp increases also, meaning a drop in the capillary perfusion pressure.

6.2.4 Ocular hypertension and glaucoma

Whereas the clinical evidence from the above mentioned examples of asymmetric DR seems to point to the general idea that an increase in end capillary perfusion pressure is a main risk for the development or worsening of DR, studies directly examining the association between MOPP and DR are rare. Many studies have assessed the relationship between retinal blood flow and DR [81–84]. Whilst a series of studies have reported that increased retinal blood flow is associated with background DR, [81, 82] pre-proliferative DR, and proliferative DR, [81] as with RVO, data on the relationship of the effect of MOPP remains inconsistent globally. Whilst some researchers have shown higher MOPP is associated with DR, macular edema, and hard exudation, [81, 85] others did not observe this association [86].

The pathophysiology of glaucoma is not completely understood. However, both diabetes and glaucoma appear to share some common risk factors and pathophysiologic similarities, including the phenomenon of neurovascular coupling (NVC). Within the retina, studies have shown that not only is vascular dilatation reduced in patients with little to no DR, but also in glaucoma patients when compared to that in healthy subjects [87–89]. These studies indicate a process whereby abnormal neurovascular coupling precedes a visible angiopathy in humans in both glaucoma and DR.

Several population-based studies have shown a positive association between ocular hypertensive disorders and diabetes mellitus [90–92] whilst some shown a negative association [93, 94]. Becker et al. showed that glaucoma decreased the incidence of DR and postulated this was because the increase in IOP lowered the transmural hydraulic pressure gradient across retinal capillaries [95].

Some groups have postulated that the reduced number of retinal ganglion cells found in glaucoma lead to a reduced ischemic drive and thus prevents DR development [96, 97]. Singal et al. showed that the mean duration of diabetes with early non-proliferative changes was maximum in patients with primary open angle glaucoma (15.8 years), followed by normal tension glaucoma (14.0 years) and then in non-glaucomatous patients (13.3 years) [96]. They also observed that advanced stages of DR changes were seen more so in the group without glaucoma. These findings, alongside those of Williams et al., suggests that not only does glaucoma delay the onset of DR, but that glaucoma also has an effect in delaying the progression of DR changes [96, 97].

The protective effect of ocular hypertension and glaucoma on DR can also be understood through the hemodynamic changes.

Increased IOP causes a decrease in ocular perfusion pressure. This directly reduces endothelial baro-damage. Also, in patients with established glaucoma with significant optic atrophy, the resultant arteriole narrowing would additionally cause a significant increase in vascular resistance and a decrease in end capillary perfusion as demonstrated through Poisseuille's equation.

6.2.5 Pan retinal laser photocoagulation (PRP)

PRP is an effective treatment against proliferative DR. The DRS study showed that the 5 year rate of severe vision loss from proliferative DR was reduced from 50% without PRP to 20% by this treatment, as onset and progression of neovascularization was prevented. Furthermore, PRP reduced the risk of elevated IOP during the study period thus delaying onset of neovascular glaucoma [98]. It's therapeutic effect is understood to be due to the reduction of oxygen and metabolic demand through tissue ablation. It should be pointed out that after PRP, the arteriole calibers are decreased significantly, resulting in a marked reduction of end capillary perfusion pressure.

6.2.6 Anti-VEGF intravitreal injections

Development of injectable anti-VEGF agents into the eye have revolutionized the way in which diabetic macular edema and proliferative DR can be managed. Multiple pivotal trials reproducibly demonstrated significant regression of DR severity with anti-VEGF treatments, [99–101] as well as complete regression of new vessels in up to 20% of cases [102]. It has also been noted that in diabetic macular edema treatment, the number of injections needed to control macula edema decreases in the second and third year as compared to the first year of treatment [99]. These findings suggest that continued anti-VEGF, with attendant restoration of healthier hemodynamics, a reduced capillary perfusion pression and less baro-damage to the capillary endothelium may confer some improvement in the severity of retinopathy.

7. Conclusions

A comprehensive review of retinal hemodynamics, the interplay between various factors such as blood pressure, vascular resistance, IOP, ocular perfusion pressure and blood flow are covered in this chapter. Changes in one of more of these factors are discussed in different disease states. Specifically, we discussed two important and frequent clinical entities; CRVO and DR. IOP plays an important role in the pathogenesis of each of these two conditions. In eyes with predispositions to venous stasis, IOP causes a further reduction in ocular perfusion and thus exacerbates the stasis. On the other hand, in diabetic microangiopathy, the endothelium is damaged, so that an increased IOP with attendant reduction in perfusion pressure actually protects the endothelium from transmural pressure related trauma.

Conflict of interest

The authors declare no conflict of interest.

Author details

Fidan Jmor and John C. Chen*
Department of Ophthalmology and Visual Sciences, McGill Academic Eye Centre, Faculty of Medicine and Health Sciences, McGill University, Montreal, Quebec, Canada

*Address all correspondence to: johnchen.mcgill@gmail.com

References

[1] Wong TY, Klein R, Klein BE, Tielsch JM, Hubbard L, Nieto FJ. Retinal microvascular abnormalities and their relationship with hypertension, cardiovascular disease, and mortality. Surv Ophthalmol. 2001; 46:59-80.

[2] Leske MC. Open-angle glaucoma—an epidemiologic overview. Ophthalmic Epidemiol. 2007;14:166-172

[3] Guidoboni G, Harris A, Cassani S, et al. Intraocular pressure, blood pressure, and retinal blood flow autoregulation: a mathematical model to clarify their relationship and clinical relevance. Invest Ophthalmol Vis Sci. 2014;55:4105-4118

[4] Kur J, Newman EA, Chan-Ling T. Cellular and physiological mechanisms underlying blood flow regulation in the retina choroid in health disease. Prog Retin Eye Res. 2012. 31(5):377-406.

[5] B. Anand-Apte, J.G. Hollyfield. Developmental Anatomy of the Retinal and Choroidal Vasculature. In Encyclopedia of the Eye. Academic Press, 2010;9-15. ISBN 9780123742032,

[6] Hughes S, Yang H, Chan-Ling T. Vascularization of the human fetal retina: roles of vasculogenesis and angiogenesis. Invest Ophthalmol Vis Sci. 2000; 41:1217-1228.

[7] Chan-Ling T, McLeod DS, Hughes S, Baxter L, Chu Y, Hasegawa T, Lutty GA. Astrocyte-endothelial cell relationships during human retinal vascular development. Invest Ophthalmol Vis Sci. 2004a; 45:2020-2032.

[8] Engerman RL. Development of the macular circulation. Invest Ophthalmol. 1976; 15:835-840.

[9] Chaine, G. and Imbs, J.L. (1997) Rappel anatomo- physiologique des particularites de la circulation oculaire . Eil et pathlogie generale. Societe Francaise d'Ophthalmologie, Flament, J. and Storck, D. eds. Paris: Masson

[10] Cholkar K, Dasari SR, Pal D, Mitra AK. Eye:anatomy, physiology and barriers to drug delivery. In: Ocular transports and receptors. Woodhead publishing Limited 2013;p1-36. DOI: 10.1533/9781908818317.1

[11] Flammer J, Konieczka K and Flammer AJ. The primary vascular dysregulation syndrome: implications for eye diseases. The EPMA Journal 2013, 4:14 http://www.epmajournal.com/content/4/1/14

[12] Linsenmeier, R.; Padnick-Silver, L. Metabolic dependence of photoreceptors on the choroid in the normal and detached retina. Invest. Ophthalmol. Vis. Sci. 2000, 41, 2678-2683.

[13] McMonnies, C.W. Hyperbaric oxygen therapy and the possibility of ocular complications or contraindications. Clin. Exp. Optom. J. Aust. Optom. Assoc. 2015, 98, 122-125.

[14] Alqawlaqa S, Flanagand JG, Sivaka JM. All roads lead to glaucoma: Induced retinal injury cascades contribute to a common neurodegenerative outcome. Experimental Eye Research. 2019, 183:88-97

[15] Robinson F, Riva CE, Grunwald JE, Petrig BL, Sinclair SH. Retinal blood flow autoregulation in response to an acute increase in blood pressure. Invest Ophthalmol Vis Sci. 1986; 27:722-726.

[16] Newman, E.A. Functional hyperemia and mechanisms of neurovascular coupling in the retinal vasculature. J. Cereb. Blood Flow Metab. 2013.

[17] Riva, C.E.; Logean, E.; Falsini, B. Visually evoked hemodynamical response and assessment of neurovascular coupling in the optic nerve and retina. Prog. Retin. Eye Res. 2005, 24, 183-215.

[18] Wareham LK and Calkins DJ. The Neurovascular Unit in Glaucomatous Neurodegeneration. Front Cell Dev Biol. 2020; 8: 452.

[19] Grunwald JE, Riva CE, Stone RA, Keates EU, Petrig BL. Retinal autoregulation in open-angle glaucoma. Ophthalmology. 1984b; 91:1690-1694.

[20] Grunwald JE, Riva CE, Kozart DM. Retinal circulation during a spontaneous rise of intraocular pressure. Br J Ophthalmol. 1988; 72:754-758.

[21] Dumskyj MJ, Eriksen JE, Dore CJ, Kohner EM. Autoregulation in the human retinal circulation: assessment using isometric exercise, laser Doppler velocimetry, and computer-assisted image analysis. Microvasc Res. 1996; 51:378-392.

[22] Harris A, Arend O, Bohnke K, Kroepfl E, Danis R, Martin B. Retinal blood flow during dynamic exercise. Graefes Arch Clin Exp Ophthalmol. 1996a; 234:440-444.

[23] Polska E, Simader C, Weigert G, Doelemeyer A, Kolodjaschna J, Scharmann O, Schmetterer L: Regulation of choroidal blood flow during combined changes in intraocular pressure and arterial blood pressure. Invest Ophthalmol Vis Sci 2007, 48:3768-3774.

[24] Neuhuber W, Schrodl F: Autonomic control of the eye and the iris. Auton Neurosci 2011, 165:67-79.

[25] Sigal IA, Flanagan JG, Tertinegg I, Ethier CR. Modeling individual-specific human optic nerve head biomechanics. Part I: IOP-induced deformations and influence of geometry. Biomech Model Mechanobiol. 2009;8:85-98.

[26] Sigal IA, Flanagan JG, Tertinegg I, Ethier CR. Modeling individual-specific human optic nerve head biomechanics. Part II: influence of material properties. Biomech Model Mechanobiol. 2009;8:99-109.

[27] Sigal IA. An applet to estimate the IOP-induced stress and strain within the optic nerve head. Invest Ophthalmol Vis Sci. 2011;52:5497-5506.

[28] Ganesan P, He S, Xu H. Development of an image-based model for capillary vasculature of retina. Comput Methods Programs Biomed. 2011;102:35-46.

[29] Takahashi T, Nagaoka T, Yanagida H, et al. A mathematical model for the distribution of hemodynamic parameters in the human retinal microvascular network. J Biorheol. 2009;23: 77-86.

[30] Findl O, Strenn K, Wolzt M, et al. Effects of changes in intraocular pressure on human ocular haemodynamics. Curr Eye Res. 1997;16:1024-1029.

[31] Galassi F, Giambene B, Corvi A, Falaschi G, Menchini U. Retrobulbar hemodynamics and corneal surface temperature in glaucoma surgery. Int Ophthalmol. 2008;28:399-405.

[32] Harris A, Joos K, Kay M, et al. Acute IOP elevation with scleral suction: effects on retrobulbar haemodynamics. Br J Ophthalmol. 1996;80:1055-1059.

[33] Cantor LB. The effect of trabeculectomy on ocular hemodynamics. Trans Am Ophthalmol Soc. 2001;99:241

[34] Harris A, Garzozi HJ, McCranor L, Rechtman E, Yung CW, Siesky B. The effect of latanoprost on ocular blood flow. Int Ophthalmol. 2009;29:19-26.

[35] Hommer A, Sperl P, Resch H, et al. A double-masked randomized crossover study comparing the effect of latanoprost/timolol and brimonidine/timolol fixed combination on intraocular pressure and ocular blood flow in patients with primary open-angle glaucoma or ocular hypertension. J Ocul Pharmacol Ther. 2012;28:569-575.

[36] Inan UU, Ermis SS, Yucel A, Ozturk F. The effects of latanoprost and brimonidine on blood flow velocity of the retrobulbar vessels: a 3-month clinical trial. Acta Ophthalmol Scand. 2003;81:155-160.

[37] Chen JC, & Cheema D: Repeated anterior chamber paracentesis for the treatment of central retinal artery occlusion. *Can J Ophthalmol*29:207-209, 1994.

[38] Cugati S, Wang JJ, Rochtchina E & Mitchell P (2006): Ten-year incidence of retinal vein occlusion in an older population: the Blue Mountains Eye Study. Arch Ophthalmol 124: 726-732.

[39] Hayreh SS. Prevalent misconceptions about acute retinal vascular occlusive disorders. Prog Retin Eye Res. 2005;24(4):493-519.

[40] Rehak J, Rehak M. Branch retinal vein occlusion: pathogenesis, visual prognosis, and treatment modalities. Curr Eye Res. 2008;33(2):111-31.

[41] Verhoeff FH. The effect of chronic glaucoma on the central retinal vessels. Arch Ophthalmol 1913; 42: 145-52.

[42] Larsson S, Nord B. Some remarks on retinal vein thrombosis and its treatment with anticoagulants. Acta Ophthalmol (Kbh) 1950; 28:187-201.

[43] Vannas S, Tarkhanan A. Retinal vein occlusion and glaucoma. Br J Ophthalmol 1960; 44: 583-9.

[44] Yin X, Li J, Zhang B, Lu P. Association of glaucoma with risk of retinal vein occlusion: A Meta-analysis. Acta Ophthalmologica 2009;97:652-9.

[45] Pasco M, Singer L, Romen M. Chronic simple glaucoma and thrombosis of retinal vein. Ear Nose Throat J 1973; 52: 294-7.

[46] Ring CP, Pearson TC, Sanders MD, Wetherley-Mein G. Viscosity and retinal vein thrombosis. Br J Ophthalmol 1976; 60: 397-410.

[47] Dodson PM, Kritzinger EE. Management of retinal vein occlusion. Br Med J 1987; 295: 1434-5.

[48] Bertelsen TI. The relationship between thrombosis in the retinal veins and primary glaucoma. Acta Ophthalmol (Copenh) 1961;39: 603– 613.

[49] Barnett EM, Fantin A, Wilson BS, Kass MA & Gordon MO. The incidence of retinal vein occlusion in the ocular hypertension treatment study. Ophthalmology 2010;117: 484-488.

[50] Thapa R, Paudyal G & Bernstein PS. Demographic characteristics, patterns and risk factors for retina vein occlusion in Nepal: a hospital-based case-control study. Clin Exp Ophthalmol 2010;38: 583-590.

[51] Park HL, Jung Y, Han K, Lee MY & Park CK. Health care claims for primary open angle glaucoma and retinal vein occlusion from an 11-year nationwide dataset. Sci Rep 2017;7(1): 8038.

[52] Hayreh SS. Progress in the understanding of the vascular etiology of glaucoma. Curr Opinion Ophthalmol 1994;5: 26-35.

[53] Riva CE. Grunwald JE, Petrig BL. Autoregulation of human retinal blood flow. Invest Ophthalmol Vis Sci 1976: 27: 1706- 12.

[54] Sonnsjo B & Krakau CET. Arguments for a vascular glaucoma

etiology. Acta Ophthalmol (Copenh) 1993;71: 433-444.

[55] Gao J, Liang Y, Wang F et al. Retinal vessels change in primary angle-closure glaucoma: the Handan Eye Study. Sci Rep 2015;5: 9585

[56] Luntz MH & Schenker HI. Retinal vascular accidents in glaucoma and ocular hypertension. Surv Ophthalmol 1980;25: 163-167

[57] Frucht J, Shapiro A & Merin S. Intraocular pressure in retinal vein occlusion. Br J Ophthalmol 1984;68: 26-28

[58] Hayreh SS, Zimmerman MB, Beri M, Podhajsky P. Intraocular Pressure abnormalities associated with central and hemicentral retinal vein occlusion. Ophthalmology. 2004;11(1):133-41.

[59] *Starling, Ernest H. (1896-05-05).* "On the Absorption of Fluids from the Connective Tissue Spaces". *The Journal of Physiology. 19 (4): 312-326.* doi:10.1113/jphysiol.1896.sp000596.

[60] Grunwald JE. Effect of topical timolol on the human retinal circulation. Invest Ophthalmol Vis Sci 1986: 27: 1713-9.

[61] Zatz R and Brenner B. Pathogenesis of diabetic microangiopathy. The American Journal of Medicine 1986;80:443-53.

[62] MacMillan D: The effect of diabetes on blood flow properties. Diabetes 1983; 32(suppl 2): 56-63.

[63] Colwell JA, Halushka PV, Sarji K, Levin J, Sage1 J, Nair RMG: Altered platelet function in diabetes mellitus. Diabetes 1976; 25: 826-831.

[64] Giugliano D, Ceriello A, Paolisso G. Oxidative stress and diabetic vascular complications. Diabetes Care. 1996;19:257-267.

[65] Ciulla TA, Amador AG, Zinman B. Diabetic retinopathy and diabetic macular edema: pathophysiology, screening, and novel therapies. Diabetes Care. 2003;26:2653-2664.

[66] Schmetterer L, Wolzt M. Ocular blood flow and associated functional deviations in diabetic retinopathy. Diabetologia. 1999;42: 387– 405

[67] Grunwald JE, Riva CE, Brucker AJ, Sinclair SH, Petrig BL. Altered retinal vascular response to 100% oxygen breathing in diabetes mellitus. Ophthalmology. 1984;91:1447-1452.

[68] Sieker HO, Hickam JB: Normal and impaired retinal vascular reactivity. Circulation. 1953: 7 (79-83).

[69] UK Prospective Diabetes Study Group: Risks of progression of retinopathy and vision loss related to tight blood pressure control in type 2 diabetes mellitus: UKPDS 69. Arch Ophthalmol 2004;122:1631-40.

[70] Mallika P, Tan A, S A, T A, Alwi SS, Intan G. Diabetic retinopathy and the effect of pregnancy. *Malays Fam Physician*. 2010;5(1):2-5. Published 2010 Apr 30.

[71] Moloney JB, Drury MI. The effect of pregnancy on the natural course of diabetic retinopathy. *Am J Ophthalmol 1982;* 93:745-756.

[72] Klein R, Klein BE, Moss SE, Davis MD, DeMets DL. The Wisconsin epidemiologic study of diabetic retinopathy. III Prevalence and risk of diabetic retinopathy when age at diagnosis is 30 or more years. *Arch Ophthalmol 1984;* 102:527-532.

[73] Chen HC, Newsom RS, Patel V, Cassar J, Mather H, Kohner EM. Retinal blood flow changes during pregnancy in women with diabetes. Invest Ophthalmol Vis Sci 1994;35:3199-3208.

[74] Brown Gc, Magargal LE, Simeone FA, Goldberg RE, Federman JL, Benson WE. Arterial obstruction and ocular neovascularization. Ophthalmology 1982;89:139--44.

[75] Duker JS, Brown Gc, Bosley TM, Colt CA, Reber R. Asymmetric proliferative diabetic retinopathy and carotid artery disease. Ophthalmology 1990;90:869-74.

[76] Sebag I. Delori Fe. Effects of optic atrophy on retinal blood flow and oxygen saturation in humans. Arch Ophthalmol 1989;107:222-6.

[77] Weiter J], Zuckerman R. The influence of the photoreceptorRPE complex on the inner retina: an explanation for the beneficial effects of photocoagulation. Ophthalmology 1980;87:1133--41.

[78] Sutera SP, Skalak R. The History of Poiseuille's law. Annual review of Fluid Mechanics. 1993;25:1-19.

[79] Dogru M, Inoue M, Nakamura M, Yamamoto M. Modifying factors related to asymmetric diabetic retinopathy. Eye (Lond). 1998;12:929-33. doi: 10.1038/eye.1998.241.

[80] Avetisov ES, Savitskaya NF. Some features of ocular microcirculation in myopia. Aim Ophthalmol 1977;9:1261--4.

[81] Patel V, Rassam S, Newsom R, et al. Retinal blood flow in diabetic retinopathy. BMJ, 1992; 305, 678–83.

[82] Dimitrova G, Kato S, Yamashita H, et al. Relation between retrobulbar circulation and progression of diabetic retinopathy. Br J Ophthalmol, 2003; 87, 622–5

[83] Konno S, Feke GT, Yoshida A, et al. Retinal blood flow changes in type I diabetes. A long-term follow-up study. Invest Ophthalmol Vis Sci, 1996; 37, 1140–8.

[84] Cuypers MH, Kasanardjo JS, Polak BC. Retinal blood flow changes in diabetic retinopathy measured with the Heidelberg scanning laser Doppler flowmeter. Graefes Arch Clin Exp Ophthalmol, 2000; 238, 935–41.

[85] Moss SE, Klein R, Klein BE. Ocular factors in the incidence and progression of diabetic retinopathy. Ophthalmology , 1994; 101, 77–83.

[86] Raman R, Gupta A, Kulothungan V, et al. Association of mean ocular perfusion pressure and diabetic retinopathy in type 2 diabetes mellitus: Sankara Nethralaya diabetic retinopathy epidemiology and molecular genetic study (SN-DREAMS, 8. Report 28). Invest Ophthalmol Vis Sci, 2011; 52, 4592–7.

[87] Garhofer, G.; Zawinka, C.; Resch, H.; Kothy, P.; Schmetterer, L.; Dorner, G.T. Reduced response of retinal vessel diameters to flicker stimulation in patients with diabetes. Br. J. Ophthalmol. 2004, 88, 887-891.

[88] Lecleire-Collet, A.; Audo, I.; Aout, M.; Girmens, J.F.; Sofroni, R.; Erginay, A.; Le Gargasson, J.F.; Mohand-Said, S.; Meas, T.; Guillausseau, P.J.; et al. Evaluation of retinal function and flicker light-induced retinal vascular response in normotensive patients with diabetes without retinopathy. Invest. Ophthalmol. Vis. Sci. 2011, 52, 2861-2867.

[89] Mandecka, A.; Dawczynski, J.; Blum, M.; Muller, N.; Kloos, C.; Wolf, G.; Vilser, W.; Hoyer, H.; Muller, U.A. Influence of flickering light on the retinal vessels in diabetic patients. Diabetes Care 2007, 30, 3048-3052.

[90] Zhao D, Cho J, Kim MH, Friedman SD, Eliseo G. Diabetes, fasting glucose, and the risk of glaucoma: A

meta-analysis. Ophthalmology. 2015;122(1):72-8.

[91] Mitchell P, Smith W, Chey T, Healey PR. Open-angle glaucoma and diabetes: The blue mountains eye study, Australia. Ophthalmology. 1997;104(4):712-8.

[92] Klein BE, Klein R, Jensen SC. Open-angle glaucoma and older-onset diabetes. The beaver dam eye study. Ophthalmology. 1994;101(7):1173-7.

[93] Suzuki Y, Iwase A, Araie M, Yamamoto T, Abe H, Shirato S, et al. Tajimi Study Group. Risk factors for open-angle glaucoma in a Japanese population: the Tajimi Study. Ophthalmology. 2006;113(9):1613-7.

[94] Ellis JD, Evans JM, Ruta DA, Baines PS, Leese G, MacDonald MT, et al. Glaucoma incidence in an unselected cohort of diabetic patient: Is diabetes mellitus a risk factor for glaucoma? DARTS/ MEMO collaboration. Diabetes audit and research in Tayside study. Medicine monitirung unit. Br J Ophthalmol. 2000;84:1218-24.

[95] Becker B: Diabetes and glaucoma. In: Kimura SJ, Caygill WM, eds. Vascular complications of diabetes mellitus with special emphasis on microangiopathy of the eye. St. Louis: CV Mosby, 1967; 657-663.

[96] Williams PD, He Y, Kooner KS, Mitchell K. The suppressive effect of glaucoma on diabetic retinopathy. Invest Ophthalmol Vis Sci. 2004;45(13):4101.

[97] Singal AK. Effect of Glaucoma on Development of Diabetic Retinopathy in Long Standing Diabetics. J Clin Ophthalmol Eye Disord. 2021; 5(1): 1033.

[98] The diabetic retinopathy study research group: Photocoagulation treatment of proliferative diabetic retinopathy: Clinical applications of Diabetic Retinopathy Study (DRS) findings. DRS Report No. 8. Ophthalmology 1981;88:583-600.

[99] Diabetic Retinopathy Clinial Research Network. Randomized trial evaluating ranibizumab plus prompt or deferred laser or triamcinolone plus prompt laser for diabetic macular edema. Ophthalmology 2010; 117(6):1064-77.

[100] Nguyen QD, Brown DM, Marcus DM et al. Ranibizumab for diabetic macular edema: Results from 2 phase III randomised trials: RISE and RIDE. Ophthalmology 2010;119(4):789-801.

[101] Brown DM, Schmidt-Erfurth U, Do DV, et al. Intravitreal aflibercept for diabetic macular edema: 100-week results from the VISTA and VIVID studies. Ophthalmology 2015; 122(10):2044-2052.

[102] Sivaprasad S, Prevost AT, Bainbridge J, et al. Clinical efficacy and mechanistic evaluation of aflibercept for proliferative diabetic retinopathy (acronym CLARITY): a multicentre phase IIb randomised active-controlled clinical trial. BMJ Open 2015;5:e008405.

Chapter 3

Progression from Ocular Hypertension into Glaucoma

Sayantan Biswas

Abstract

Ocular hypertension (OHT) is characterized by raised intraocular pressure (IOP) >21 mmHg without any visual field (functional) or optic nerve (structural) defect featuring glaucoma. Raised IOP is a major risk factor of glaucoma and a proportion of eyes with OHT progresses into primary open angle glaucoma. Glaucoma is a debilitating disease with potential for blindness if left untreated and associated reduction in the quality of life of the affected individual. It is challenging for the clinicians to decide whether an OHT will progress into glaucoma or not based on the risk factor model of the Ocular hypertension treatment study. Moreover, the question whether only IOP or a myriad of factors like central corneal thickness, baseline IOP, visual field, family history of glaucoma, ocular biomechanics are all important in determining the progression is yet to be answered. The rate of progression is also important and needs analysis for further discussion. Summarizing the landmark studies on ocular hypertension and glaucoma to date are imperative in this regard. This chapter presents the overview of OHT and its possible etiology and pathophysiology, risk factors, clinical tests evaluating OHT eyes and elaborates on the progression of OHT to glaucoma over time in relation to the treatment.

Keywords: Ocular hypertension, pathophysiology, risk factors, progression, treatment

1. Introduction

Elevated intraocular pressure (IOP) is an important and the only clinically modifiable risk factor of glaucomatous optic neuropathy [1–4]. Clinically, ocular hypertension (OHT) is most widely defined as raised IOP of > 21 [5] or ≥24 [6] mm Hg in eyes without detectable glaucomatous visual field (VF), optic nerve (ON) or retinal nerve fiber layer (RNFL) damage [6, 7]. It may be also defined as IOP in the highest 97.5% percentile for the population without having optic disc or visual field damage [8]. The prevalence of OHT worldwide varies between 0.32-12.2% [9–15].

Ocular hypertensives have been studied widely through several clinic and population-based studies. The two earliest prospective studies are done by Quigley et al [2] and Georgopoulos et al [16] on large cohort of OHT patients. The first one is the follow-up of 647 OHT patients with IOP >21 mm Hg for 6.2 years. VF testing using Goldmann perimeter (static & kinetic) was performed yearly once [2]. The other study was on 345 untreated patients with OHT (IOP≥21 mm Hg) for a mean duration of 7.3 (6-8) years. VF testing was done every 6-10 months using Humphrey VF analyzer 30-2 program [16].

Ocular hypertension treatment study (OHTS) is the largest multicenter randomized trial involving 1636 OHT participants (aged 40-80 years, IOP 24-32 mm Hg) randomized to either ocular hypotensive drug or to stay under observation for a follow up period of 7.5 years. They were followed up with Humphrey VF 30-2 every 6 months and stereoscopic optic disc photographs every 12 months. Treatment goal was IOP reduction by 20% or more and an IOP of ≤24 mm Hg. The average reduction of IOP achieved with medication was 18.4% which is equivalent to 4.6 mm Hg. In the second phase of the study, the two groups of medication and observation was treated with IOP lowering drugs for another 5.5 years (total 13 years), creating an early treatment and delayed treatment groups [6, 7, 17].

Most population-based studies have shown 9.5-17.4% of OHT eyes develop primary open angle glaucoma (POAG) without treatment over 5 years [6, 17, 18]. Around 1.5-10.5% of these untreated OHT eyes stand at a risk of going blind over the next 15 years [19]. Even after treatment, 4.4% of OHT eyes develop POAG [17] while 0.3-2.4% eyes still carry the risk of becoming blind [19]. Hence, OHT poses a significant socio-economic impact with potential to debilitate the quality of life (QoL) of the diagnosed person once the disease starts progressing [20–24].

2. Pathophysiology

Lamina cribrosa (LC) is a sieve-like structure, which creates a perforation in the sclera through which the axons of the retinal ganglion cells (RGC) exit the eye as optic nerve fibers. The lamina is the weakest point in the wall of the eye. IOP is believed to cause mechanical stress and strain on the LC at the posterior structures of the eye [25]. This results in the compression, deformation and remodeling of the LC at the optic nerve head (ONH) along with axonal damage and blockade of RGC axonal transport (both orthograde and retrograde) to the lateral geniculate nucleus (LGN) [26–28]. It is followed by apoptotic degeneration and death of RGC of the retina and the optic nerve causing vision loss [29].

In experimental glaucoma, disruption of axonal transport causes the collections of vesicles and disorganization of microtubules and neurofilaments in the prelaminar and postlaminar regions [29]. Postmortem of human eyes with glaucoma also revealed these ultrastructural changes in the optic nerve fibers [25]. It was shown that the pressure gradient between the intraocular pressure and cerebral spinal fluid pressure at the lamina cribrosa might be influential in maintaining blood flow in the optic nerve head [30]. When the pressure gradient is increased, the axonal transport would be disrupted, leading to retinal ganglion cell damage. In fact, the degree of RGC death is related to the level and duration of intraocular pressure elevation [29].

This strain also initiates a cascade of molecular and neurotransmitter changes in the surrounding cells of the retina and optic nerve (like astrocytes, microglia, horizontal and amacrine cells, etc.) which alters the microcirculation and remodels the extracellular matrix [29]. Further atrophy and death of the target relay neurons occurs in the magnocellular and parvocellular LGN [28].

The intraocular pressure is a function between the production of aqueous humor from the ciliary processes of the ciliary body and its outflow through the trabecular meshwork (TMW) via schlemm's canal (conventional pathway) and the uveoscleral pathway via ciliary muscle/choroid/sclera (unconventional pathway) [29, 31]. There is an increased resistance found in the aqueous humor outflow through the TM lead conventional pathway in OHT eyes [29, 32]. Thus, resulting in an increase in the IOP, which causes the mechanical stress and strain on the posterior structure of the eye as described.

The average IOP which results from a balance between the normal production and outflow of aqueous humor, which is around 14-15 mm Hg [31, 33]. Although, it is almost impossible to define what is a normal or safe IOP as all individual eyes are uniquely susceptible to the damage caused by IOP. Eyes do not develop any glaucomatous damage in spite of relatively high IOP, whereas, others get damaged even under normal or even relatively low IOP [34, 35]. The result of genetic predisposition and risk factors working alone or their interactions along with the biomechanical properties of the ONH and the scleral connective tissue are hypothesized to account for the susceptibility of individuals under high, normal or low IOP [26].

Secondary damage may occur consequential to RGC death due to the release of glutamate and glycine from the injured neurons leading to excitotoxic damage [36, 37]. Production of nitric oxide may result in oxidative damage to the RGCs and their axons [38–42]. Tissue ischemia-hypoxia is another implicated factor related to glaucomatous optic neuropathy [43, 44]. Reduced ocular perfusion pressure, which is dependent on the systolic and diastolic blood pressure is also found to be associated with higher incidence of OAG [45–47]. Other causes are vascular insufficiency and autonomic dysfunction of the ONH [48–51].

3. Risk factors

3.1 Age

Age is major factor positively associated with the IOP [52]. The Los Angeles Latino Eye Study found higher prevalence of OHT among older Latinos than in younger Latinos ($P < 0.0001$) [53]. Latinos aged ≥80 years had a 3-times higher prevalence than the younger ones (40-49 years) [10]. The OHTS [6] and European Glaucoma Prevention Study (EGPS) [54] found older age (per decade) to have higher risk of progressing from OHT into POAG (hazard ratio (HR) 1.22, $P < 0.05$ and 1.32, $P < 0.05$ respectively). Similarly, a 6 year follow up of urban Australian patients with POAG & OHT to show a significant association of age with the prevalence of POAG [55] and Malmo¨ Ocular Hypertension Study (MOHS) [56] also found older age (per year) as a predictive factor (HR 1.32, $P < 0.05$ and 1.05, $P = 0.034$) of developing POAG among OHT patients in their multivariate analysis.

3.2 Intraocular pressure

IOP is the strongest risk factor associated with glaucoma such that it is regarded as causality. The dose-response relationship has been well documented and demonstrated in several prevalence and longitudinal studies [7, 57–59]. OHTS demonstrated that 23% reduction in the IOP can decrease the incidence of POAG by 60% [7]. Similarly, the Melbourne Visual Impairment Project estimated that for every 1 mm Hg, the risk for glaucoma increased by 10% [60]. Also, the Early manifest Glaucoma Trial (EMGT) and the Collaborative Normal Tension Glaucoma Study (CNTGS) reported an IOP reduction of 25% and greater than 30% can lower the risk of progression by 33% and 50%, respectively, compared to those with no treatment [61, 62]. The Advanced Glaucoma Intervention Study (AGIS) also reported a reduction of IOP to be associated with stable visual fields [57]. A retrospective cohort analysis of 230 OHT patients over 5 years revealed that higher peak IOP is a risk factor for developing POAG in the multivariate analysis. Both the peak IOP and the mean IOP in the progressed group was higher than in the stable group ($P < 0.01$) [63]. IOP per mmHg presented with HR of 1.14 ($P = 0.047$) for developing POAG among OHT eyes in MOHS [56].

3.3 Central corneal thickness

OHTS confirmed thinner central corneal thickness (CCT) to be associated with greater risk of conversion into POAG from OHT [6]. The risk increased by 71% for every 40 μm decrease in the CCT (Multivariate HR 1.71, $P < 0.05$). Similarly, the EGPS found lower CCT by 40 μm to have higher risk (HR 1.32, $P = 0.018$) of POAG [54]. Eyes with thickness ≤555 μm had 3 times increased risk of developing POAG than those with CCT >580 μm [6]. This was probably because thicker cornea has an actual (true) IOP which is lesser than the measured IOP. Conversely, thinner cornea has a true IOP which is higher than measured IOP. Thus, eyes with thicker cornea stand at a risk of getting misdiagnosed as OHT. However, we do not know whether the corneal thickness is associated with factors affecting susceptibility to glaucoma or not.

3.4 Corneal parameters and intraocular pressure and

Corneal properties such as thickness, astigmatism, curvature, hysteresis and biomechanics poses a challenge in measuring the true IOP [64–66]. The Goldmann applanation tonometer (GAT) is known to falsely elevate IOP in thick cornea and falsely reduce IOP in thin corneas [67, 68]. The IOP values measured using CorVis ST is shown to remain almost unaffected by corneal parameters like its thickness and topography through a wide range of IOPs. CorVis ST IOPs were validated on ex-vivo human donor eyes [69, 70]. Corneal characteristics are believed to be strong confounding factors in the measurement of true IOP [70].

3.5 Race

In Phase 2 of OHTS, POAG developed more commonly among African-Americans in the univariate analysis but loses its significance on adding vertical cup to disc ratio (VCDR) and CCT into the multivariate model [6, 17]. However, with similar baseline IOP, follow up IOP and treatment, African Americans have higher risk of developing POAG. This suggests that black race is not associated with an increased risk of glaucoma progression. However, the higher prevalence of other risk factors of glaucoma is present in black individuals such as thinner central corneal thickness (CCT), higher IOP, larger VCDR than their white counterparts [71–73]. Self-reported black race was also identified as an independent risk factor of developing optic disc hemorrhage after 13 years of follow up in OHTS [74].

3.6 Gender

Although males with OHT presented with a higher risk of POAG in the univariate analysis of OHTS (OR 1.87, $P<0.05$), it was not significant in the multivariate model [6]. However, other studies on OHT & POAG patients found male sex to have higher odds of having POAG than females (OR 1.9, $P <0.01$) [55].

3.7 Family history

OHTS failed to find any association of OHT progression with family history of glaucoma [6]. On the other hand, Landers et al studied 301 OHT and 438 POAG patients and reported family history of glaucoma to be a risk factor of having POAG (Odds ratio 1.6, $P < 0.01$) [55]. Similarly, an earlier study on 345 OHT patients revealed that out of 31.6% with family history of glaucoma, 55% developed POAG ($P < 0.001$), which shows family history (heredity) as an important factor in the development of glaucoma [16].

3.8 Myopia

OHTS have found no association between myopia and POAG [6]. However, earlier studies involving patients with OHT and myopia were found to develop glaucoma more than those without myopia [75, 76]. Landers et al. studied patients with POAG (n = 438) and OHT (n = 301) with SAP for a duration of 6 years and reported that myopic patients (SE ≤ -1 D) with OHT have 1.5 times higher risk of developing POAG [55]. Georgopoulos et al. studied 345 untreated OHT with SAP over a period of 7.3 years and found axial myopia (0.001 < P < 0.01) to be a risk factor for the development of glaucoma [16]. Similarly, the Casteldaccia Eye Study on 44 OHT/POAG (IOP ≥ 24 mmHg) and 220 controls (IOP ≤ 20 mm Hg) found myopia to be associated with increased (multivariate OR 5.56) of OHT/POAG [77].

Quigley et al. followed 647 OHT patients (40% under treatment) with refractive errors in the range between +12 and -12 D for a period of 6.2 years with Goldmann kinetic perimeter. They showed that there was no association between refractive error and visual field progression [2].

Similarly, the Malmo¨ Ocular Hypertension Study randomized 90 OHT patients to topical timolol or placebo and observed every 3 months for 10 years to conclude that myopia have no influence on the visual field progression. In their cohort, 35% of the myopes progressed compared to the 54% of non-myopes, and there was no significant association between refractive error and VF loss in OHT patients [56].

3.9 Optic disc hemorrhage

Optic disc hemorrhage (ODH) in OHT eyes was associated with a 3.7 times higher risk of developing into POAG (P < 0.001) in multivariate model which included the baseline factors predictive of POAG. The incidence of POAG in eyes with and without ODH were 13.6% and 5.2% respectively after 8 years of follow up of OHTS [78]. European Glaucoma Prevention Study (EGPS) also established ODH as an independent risk factor of POAG with HR of 1.97 [79]. After the end of the 13 years follow of OHTS, it was further confirmed in the multivariate analysis that ODH has a 2.6-fold increased risk of converting to POAG (P < 0.0001) [74].

4. Visual field testing

Standard automated white-on-white perimetry (SAP) is the most extensively investigated tool for assessment of visual field defects in glaucoma. In SAP, visual sensitivities at pre-defined locations of the retina are measured and compared with age-corrected normative values to detect locations of abnormal visual field sensitivity. SAP employs white stimuli on a white background to quantify visual sensitivity [80]. Visual field parameters including the mean deviation (MD), pattern standard deviation (PSD) and visual field index (VFI) [81] are global indices for measurement of average visual sensitivity and function of an eye. MD is calculated by weighting and averaging the differences of sensitivity thresholds for all the tested points between a subject's thresholds and the normative values. A negative MD indicates overall depression in visual sensitivity. PSD is an index indicating the uniformity of visual field sensitivity. It is determined by comparing the differences between adjacent points. A high PSD value indicates focal visual field loss. A low PSD value, however, can be found in a normal visual field or in an eye with diffuse loss in visual sensitivity. VFI is a percentage of overall visual field sensitivity compared with the normal age-adjusted visual field. VFI has been shown to be less influenced by cataract compared with MD [81]. The severity of glaucomatous damage can be classified into mild (MD ≥ -6 dB) and moderate-to-advanced (MD <-6 dB) according to the Hodapp-Parrish-Anderson criterion [82]. Glaucoma hemifield test (GHT) [83] is another important parameter incorporating into SAP for glaucoma detection. GHT is calculated by the comparison of five clusters of corresponding test points between the superior and inferior fields. GHT reports the asymmetry of visual field defects in glaucoma. Three categories of visual field are classified by GHT: "within normal limits", "borderline" and "outside normal limits". Yet, a "within normal limits" visual field does not always represent a normal field. Although SAP has been proven to be useful for detection and monitoring of glaucoma [84], early stages of glaucomatous damage may appear as normal in SAP. Because of specific visual function of retinal ganglion cells (RGC) subtypes, selective perimetry can isolate the specific RGCs populations, which was found to detect glaucoma earlier than SAP. There are mainly two types of function-specific perimetries, short-wavelength automated perimetry (SWAP) and frequency doubling technology (FDT). SWAP selectively tests RGCs that target the koniocellular sublayers of the lateral geniculate nucleus by projecting blue stimulus on yellow background. In longitudinal studies, it can detect glaucoma as early as 5 years compared to standard perimetry [85–87]. FDP tests large diameter retinal ganglion cells that target magnocellular layers of the lateral geniculate nucleus and can also detect glaucoma earlier [85, 86, 88]. In a study comparing the diagnostic capability among SAP, SWAP and FDT for detection of early glaucoma (MD > -6 dB), the sensitivities were 46%, 34% and 52% respectively, with specificity ≥97% [89].

In clinical practice, SAP remains the most widely used visual field assessment for diagnosing and monitoring glaucoma. Detecting VF progression (change) over time is difficult and challenging owing to the fact that VF result are largely influenced by several factors [90–95]. Most studies formulated their own criteria to detect VF change with their own merits and demerits [96].

4.1 Visual field testing in ocular hypertensives

Although, optic disc photographs detected most (55%) of the early glaucomatous changes, almost one third had visual field changes as their earliest glaucomatous change in OHTS [7]. Soliman et al evaluated the diagnostic sensitivity of standard automated perimetry (SAP), frequency doubling technology (FDT) perimetry (C-30 full threshold) and short wavelength automated perimetry (SWAP) for the detection of glaucoma damage [97]. The diagnostic performance among FDT perimetry, SWAP and SAP were compared in 42 patients with early to moderate glaucoma, 34 with ocular hypertensives, 22 glaucoma suspects, and 25 normal controls. They found that FDT had similar sensitivity in detecting visual field abnormality compared with SAP but SWAP had a poorer performance in distinguishing the normal group from glaucoma group. The study outcomes were based on measurements of MD, PSD, and percentage of abnormal points. In glaucoma patients, whose baseline SAP was abnormal, FDT perimetry and SAP detected more abnormal points than SWAP. FDT perimetry detected larger defects in ocular hypertension and glaucoma suspects, who showed a normal baseline SAP.

Johnson et al compared automated perimetry and SWAP in a group of ocular hypertension patients and found that SWAP deficits represent early glaucomatous damage and may be related to early changes that occur at the optic nerve head [98].

Bengtsson and Heijl compared the ability of SITA SWAP, full threshold SWAP and SAP (SITA Fast) in patients with ocular hypertension, suspicion of glaucoma (glaucomatous optic disc changes but found no evidence of visual field defect on SITA standard 30-2 SAP) and early manifest glaucoma subjects (repeatable visual field loss on GHT results) [99]. No significant difference was found between the three algorithms in detecting glaucomatous visual field abnormality. SITA SWAP was able to identify as much visual field loss as the full threshold SWAP but with a considerable reduction of test time.

In OHTS, global and localized rates of VF change were calculated from 780 eyes of 432 OHT patients based on linear regression between MD and time, and between threshold sensitivity values for each test location and time, respectively. It was noted that both the global and localized rates of VF decreased significantly ($P < 0.01$) over a mean of 14 years [100]. Pattern standard deviation or PSD is a weighted standard deviation of the differences between the measured and normal reference visual field at each test location. A high value represents irregularity which can be both due to focal loss in the VF or variability in the patient's responses. Hence, the use of PSD in OHTS has been criticized as it is highly variable and may add a source of error for the baseline as well as the follow up VF measurements [101].

5. Risk factors of visual field progression in OHT

Elevation of IOP has been consistently demonstrated in major clinical trials as a key risk factor for both the development and progression of glaucoma. Baseline IOP, average IOP during follow-up, and fluctuation of IOP has all been reported to be associated with glaucomatous visual field deterioration [6, 102]. Five baseline factors namely, older age, higher IOP, thinner central corneal thickness, larger VCDR and higher visual field pattern standard deviation (PSD) had greater risk of conversion from OHT to POAG [6, 17]. This model was reconfirmed and validated by two independent study population of the EGPS and Diagnostic Innovations in Glaucoma Study (DIGS) [103, 104]. Disc hemorrhage is another important risk factor of visual field progression in ocular hypertension and glaucoma patients [59, 79, 105]. The role of central cornea thickness in glaucoma progression is controversial. EMGT suggests an increased risk of progression in patients with a thinner CCT [106, 107]. Central corneal thickness is a known risk factors of visual field progression in patients with ocular hypertension with 70% increase in risk with every 40μm decrease in corneal thickness [6]. However, as shown in other studies, the association of CCT with visual field progression may not be significant [108, 109].

Other potential risk factors of visual field progression in glaucoma include age, bilaterality, exfoliation, lower systolic perfusion pressure and blood pressure [110–112].

6. Assessment of visual field (functional) progression in OHT

6.1 Trend-based analysis

Trend-based analysis has been used to detect localized and global loss in visual field. MD and visual field index (VFI) are global indices that have been used to estimate the overall rate of visual field progression. The pointwise linear regression (PLR) was first introduced to evaluate visual field progression by Fitzke et al. where the luminance sensitivity of every location from the entire visual field within a series of examination against time was analyzed. PLR has been shown to have a good agreement with Humphrey STATPAC-2 (glaucoma change probability analysis) in separating progressive from stable retinal locations (Kappa = 0.62) [113].

In OHTS, the global and localized VF change rates of 780 eyes from 432 OHT patients over a period of 13 years were calculated based on linear regression between MD and time, and between threshold sensitivity values for each test location and time, respectively. The significant decrease of both the global (MD) and localized rates of VF was recorded ($P < 0.01$). The predetermined criteria of -0.5 dB/year were met in 18.1% eyes. The rate of VF progression before and after the initiation of treatment was -0.23 vs. -0.06 dB/year [100]. The mean rate of change of MD was -0.08 ± 0.20, -0.26 ± 0.36 and -0.05 ± 0.14 dB/year for all, POAG and non-POAG eyes in OHTS ($P < 0.001$) [96].

6.2 Event-based analysis

The pattern deviation map and the total deviation map have been used to detect visual field progression in clinical practice and in glaucoma clinical trials. In the Early Manifest Glaucoma Trial (EMGT), visual field progression, measured by event-based analysis, was the primary outcome measurement of the study. The frequency of visual field progression was compared in 255 early glaucoma patients with and without treatment (treated with trabeculoplasty and betaxolol hydrochloride eye drops). To determine visual field progression, the follow-up visual fields were compared with the average of 2 baseline visual fields in the same eye using glaucoma change probability maps (GCPMs). GCPMs detects significant visual sensitivity worsening at $P < 0.05$ at each of 76 test point locations in the visual field. The EMGT uses pattern deviation GCPMs, rather than the standard total deviation GCPMs, to limit the impact of generalized loss in visual sensitivity secondary to cataract. Definite EMGT visual field progression was defined as at least 3 test points showing significant progression, as compared with the baseline, at the same locations on 2 consecutive GCPMs [62, 114]. The EMGT criteria have been incorporated into the Humphrey Field Analyzer Guided Progression Analysis (GPA, Carl Zeiss Meditec, Dubin, CA). The EMGT criteria have been reported to identify progression earlier than the visual field progression criteria used in AGIS and Collaborative Initial Glaucoma Treatment Study (CIGTS) [84].

7. Assessment of structural progression in OHT

HRT I was used in the Confocal Scanning Laser Ophthalmoscopy Ancillary Study to OHTS which included 865 eyes from 438 participants with ocular hypertension. Forty-one eyes from 36 participants developed POAG based on confirmed visual field defect or optic disc glaucomatous change with a median follow-up time of 48.4 months. Several baseline topographic optic disc measurements taken by HRT I were significantly associated with the development of POAG in both univariate and multivariate analyses, including cup-disc area ratio, mean cup depth, mean height contour, cup volume, reference plane height, and smaller rim area, rim area to disc area, and rim volume. In addition, the classification "outside normal limit" by Moorfields regression analysis (MRA) was also associated with POAG development. It was suggested that HRT I was useful in glaucoma prediction and can be used in glaucoma progression monitoring [115]. The same research group further compared the performance between the baseline glaucoma probability score (GPS) and MRA on predictive ability of conversion from OHT to POAG. Sixty-four eyes of 50 OHT subjects converted to glaucoma based on repeatable visual field defect or optic disc change with a median follow-up time of 72.3 months. In a multivariate analysis, "outside normal limits" global and sectoral baseline GPS showed significant association with the development of POAG with HR ranging from 2.92

to 3.70. In addition, baseline MRA parameters also showed significant association with POAG development with HR ranging from 2.41 to 11.03. It was concluded that both GPS and MRA showed similar performance in predicting the conversion of glaucoma from OHT subjects [116].

In the study by Strouthidis et al on 198 OHT and 21 normal subjects, rim area (RA) progression was calculated with linear regression of sectoral RA/time, defined as slope>1%/year; visual field progression was calculated by PLR of sensitivity/time. The specificities of RA were estimated from 88.1% to 90.5% with the less-stringent criteria, which were as high as the specificities of visual filed progression (85.7% to 95.4%) when standard criteria were used, indicating that both VF and HRT RA trend analysis had relative high specificities for glaucoma progression detection [117]. Although rim area has been shown to be useful for evaluating glaucoma progression, the agreement between rim area progression and visual field progression was often poor [118].

Event-based analysis is also useful for evaluation of rim area progression. In the study by Fayers et al, rim area change was determined using the rim area repeatability coefficient. The specificities were between 76.2% and 100% using different criteria to define rim area progression in 21 normal subjects. One hundred ninety-eight ocular hypertensive subjects were enrolled, 16.2%-45.4% of them were identified to have rim area progression based on different criteria with event-based analysis and 12% showed rim area progression based on trend-based analysis. To evaluate rim area progression, event-based analysis showed a higher progression detection rate than trend-based analysis [119].

GPS has also been used to evaluate glaucoma optic disc progression. Examining the linear regression analysis between GPS and time in 198 OHT subjects, Strouthidis et al showed that 25 subjects (12.6%) progressed by GPS with a significant negative slope ($P<0.05$); 11 of them (5.6%) also showed progression by VF with PLR analysis [120]. Twenty-six subjects (13.1%) had visual field progression alone. The specificity of GPS for glaucoma progression ranged from 95.2% to 96.8%. The conclusion is that the global GPS progression algorithm performs at least as well as previously described rim area-based HRT progression analysis [120]. Higher baseline GPS has been shown to be a risk factor for glaucoma progression in the study by Alencar et al. Two hundred and twenty-three patients with suspected glaucoma were included and followed up for an average of 63.3 months. Fifty-four (24.2%) eyes converted to glaucoma based on repeatable visual field defects and/or optic disc deterioration. Both higher values of global GPS and subjective stereophotograph assessment (larger cup-disc ratio and glaucomatous grading) were predictive of conversion, the adjusted HRs were 1.31 for global GPS, 1.34 for CDR, and 2.34 for abnormal grading, respectively. GPS performed as well as subjective assessment of optic disc in predicting glaucoma progression [121].

8. Features of optic nerve head progression

Using serial optic disc photographs of 259 patients with elevated IOP followed over 15 years, Pederson et al showed that progressive enlargement of the optic cup was commonly the first sign of glaucoma progression [122]. Tuulonen et al detected equal numbers of glaucomatous eyes with diffuse and localized enlargement of optic disc cup in 61 patients with ocular hypertension [123]. In the study by Odberg et al, progressive optic disc cupping occurred most frequently in the superotemporal or inferotemporal quadrants [124]. Lloyd et al examined serial optic disc photographs of 336 eyes of 168 patients with ocular hypertension or early glaucoma [125]. Optic disc progression was defined as: new or increased neuroretinal rim (NRR) narrowing (2 or more clock hours), notching (1 clock hour or less of narrowing of the NRR), optic disc excavation (undermining of the NRR or disc margin), or development of nerve fiber layer defect. Ninety two of 336 eyes (27.4%) showed optic disc progression after a median of 6.1 years of follow-up. Among those with progression, excavation occurred most commonly (89% of eyes), followed by rim narrowing (54% of eyes) and notching (16%). The inferotemporal quadrant of optic nerve head was the most common location for glaucoma progression [125].

Expansion in the size of peripapillary atrophy was related to glaucoma progression and conversion from ocular hypertension to glaucoma [126, 127].

Another important sign relevant to glaucoma assessment is peripapillary atrophy (PPA). The α zone PPA is located peripherally and characterized by irregular hypopigmentation and hyperpigmentation in the retinal pigment epithelium whereas β zone PPA is close to the optic disc border and characterized by a complete loss of retinal pigment epithelium [128]. Both normal subjects and glaucoma patients can develop α zone and β zone atrophy but the PPA, especially β zone PPA, is larger and more common in glaucoma patients [129]. PPA has been shown to be highly correlated to glaucomatous change [129] and its expansion is also related to glaucoma progression and conversion from ocular hypertension to glaucoma [126].

9. Features of optic nerve head changes

ONH cupping can be due to a combination of NRR loss, lamina cribrosa deformation and prelaminar surface tissue loss. Localized NRR loss can be observed as narrowing or notching of the rim, which is most frequently found in the infero-temporal and supero-temporal sectors of the optic disc [25, 130]. Progressive enlargement of the optic cup is an important sign of glaucoma progression, as shown in a study examining serial optic disc photographs in 259 patients with elevated IOP followed over 15 years [122]. Tuulonen et al studied 61 patients with ocular hypertension and reported an equal number of eyes with diffuse and localized progressive enlargement of the optic disc cup in eyes developed glaucoma over the follow up period of 10 years [123]. Odberg et al reported the supero-temporal and infero-temporal quadrants as the most frequent locations of progressive optic disc cupping [124]. Lloyd et al studied 336 eyes of 168 patients with ocular hypertension or early glaucoma for a median of 6.1 years of follow-up. They examined serial optic disc photographs and showed that for the 92 eyes (27.4%) showing optic disc progression, optic disc excavation occurred most commonly (89% of eyes), followed by rim narrowing (54% of eyes) and notching (16%). The inferotemporal quadrant of optic nerve head was the most common location for glaucoma progression [125]. In OHTS, 69 eyes had optic nerve damage alone without visual field changes. This included 55% of patients reaching the study endpoint [7]. Airaksinen et al [131] followed up 75 OHT patients for 5-15 (mean 10) years using a computerized planimeter and found a decrease in the rim area among 57% of OHT patients. Loss of RA per year was 0.47% and 2.75% among stable and progressing OHT respectively. The loss of RA was linear in half of the patients (49%), with rest as episodic (22%) and curvilinear (29%) [131].

However, it should be remembered that both OHT and EGPS used serial optic disc stereophotographs to measure the vertical CDR which is known to have high intra- and inter-observer variability among clinicians [132]. Moreover, the current optic disc imaging and measurement techniques namely HRT & OCT neither correlate well with stereophotographs and nor with the disc margin (which coincides with the Bruch's membrane opening or BMO) [133–135]. This disagreement in the VCDR measurement might give rise to variable result of OHT progression into

POAG [136]. Although, BM is a stable ONH structure and is less affected by age or increasing IOP, eyes with high myopia often have poorly visible BMO and overhanging BMO due to the BM shift [137–139].

10. Progression of ocular hypertension

Overall, 90% to 95% of patients with ocular hypertension did not go on to develop Glaucoma. After 5 years, 4.4% and 9.5% developed POAG under the medication and observation group respectively in phase 1 of OHTS. Use of medication was protective against POAG with a HR of 0.40 (95% Confidence Interval (CI): 0.27-0.59, P < 0.0001) compared to those under observation. The treatment had significant effect on both the optic disc and visual field changes. Early treatment of OHT reduces the 5-year incidence of POAG by 60% [6, 17].

In the 2nd phase of OHTS, the two groups of medication and observation were both treated with IOP lowering drugs for another 5.5 years (total 13 years), creating an early treatment and delayed treatment groups. There is a linear risk of OHT converting to POAG over 15 years. Cumulative proportion of study participants developing POAG was 0.16 vs 0.22 (P = 0.009) in early treatment vs delayed treatment groups. The median time to develop POAG was delayed in the early treatment groups than the delayed ones (8.7 vs 6 years) [17, 74, 140].

Starting treatment after the appearance of early signs of POAG do not have any significant negative effect on visual field loss over next 5 years, given the patients follow up regularly. Clinicians need to assess both the structural and functional parameters in OHT eyes to determine disease status and progression [6, 7, 17].

DIGS also found the predictive model suggested by OHTS to be useful to assess the 5-year risk of developing POAG among their independent population of 126 OHT patients [104]. DIGS also found long term IOP fluctuation not associated with risk of developing POAG in untreated OHT subjects (multivariate HR 1.08, P = 0.62). However, mean IOP, i.e., the level of IOP during follow up was significantly associated (HR 1.20 per 1mm higher, P = 0.005) [141]. Similarly, MOHS with 10-17 years of follow up of their OHT patients also found mean IOP level (HR 1.21, P = 0.005) to be significantly associated with increased risk of POAG, but not IOP fluctuation (P = 0.49) [142].

11. Treatment

OHT and POAG treatments are mainly focused at using topical prostaglandin analogs to increase the uveoscleral outflow pathway [143, 144]. Prostaglandin analogs are quite potent IOP lowering drugs which are well tolerated without much side effects and require only one dose at night to cover the nighttime peak IOP hours [144]. However, some patients still require adjunctive therapy with other drugs as topical beta-adrenergic antagonists (suppresses aqueous production), alpha-adrenergic agonists (suppresses aqueous production + increases uveoscleral outflow), and carbonic anhydrase inhibitors (suppresses aqueous production) [143, 144].

Clinical trials on topical hypotensive drops showed Latanoprostene bunod 0.024% to have a significantly better IOP lowering effect compared to either latanoprost 0.005% or timolol 0.5% over 1 year among European, North American and Japanese patients with OHT/POAG [145, 146]. The side effect profiles were similar among the medications [144].

A multicenter randomized controlled trial (Laser in Glaucoma and Ocular Hypertension trial or LiGHT) compared ocular hypotensive drops (588 eyes) vs selective laser trabeculoplasty (SLT) (590 eyes) in newly diagnosed OHT and POAG patients. This RCT was unique in its novel approach of including QOL as an outcome measure and defining the target IOP, which was specific for each individual based on their disease severity and risk of vision loss. Moreover, target IOP was adjustable based on IOP control and disease control which resembles common clinical practice more than a fixed algorithm [147, 148]. It was found that eyes treated with medicine first progressed faster compared to the laser first eyes. Total deviation (both pointwise & global) as well as pattern deviation had a greater risk of progression (risk ratio 1.37-1.55, $P < 0.001$) in the medicine first group than the laser [149]. The efficacy between SLT (611 eyes) and topical hypotensive drug (622 eyes) among eyes with OHT and POAG over 3 years were assessed through another randomized controlled trial. They found both SLT and medication to be equally effective in lowering absolute IOP in both OHT and POAG eyes. IOP control without eye drops was achieved in 75% of eyes after 1-2 SLTs [150].

Benefits of early treatment are more in high-risk patients (determined using the five-factor model of age, IOP, CCT, CDR and visual field PSD) than those with low risk. No benefit of early treatment was found for patient in the low-risk group [17].

EGPS failed to find any significant difference in IOP among OHT patients with and without dorzolamide after 5 years. Dorzolamide reduced IOP by 15-22%, whereas, the IOP in placebo group also got reduced by 9-19% ($HR < 1$, $P > 0.05$) [103].

12. Conclusion

There are several risk factors associated with the progression of OHT into glaucoma. Measurement of the true IOP is an important aspect of distinguishing patients with OHT from those with normal IOP and thicker cornea. Patients with OHT must be first evaluated and classified as having high risk or low risk of glaucoma. Only high risk OHT eyes poses a major threat of progressing into glaucoma. However, there are inherent limitations which must be considered while using the five-factor model. A better formulation of the risk assessment technique is warranted for more practical classification of OHT in clinics. Treatment with ocular hypotensive drugs or laser on high-risk patients is an effective way to reduce the risk of OHT eyes progressing into glaucoma.

Author details

Sayantan Biswas
Singapore Eye Research Institute, Singapore

*Address all correspondence to: sayantanbiswas@gmail.com

References

[1] Leidl MC, Choi CJ, Syed ZA, et al. Intraocular pressure fluctuation and glaucoma progression: what do we know? British Journal of Ophthalmology 2014;98(10):1315-9.

[2] Quigley HA, Enger C, Katz J, et al. Risk factors for the development of glaucomatous visual field loss in ocular hypertension. Archives of ophthalmo logy 1994;112(5):644-9.

[3] Leske MC, Connell AMS, Wu S-Y, et al. Risk factors for open-angle glaucoma: the Barbados Eye Study. Archives of ophthalmology 1995; 113(7):918-24.

[4] Tielsch JM, Katz J, Sommer A, et al. Hypertension, perfusion pressure, and primary open-angle glaucoma: a population-based assessment. Archives of ophthalmology 1995;113(2):216-21.

[5] Leske MC. The epidemiology of open-angle glaucoma: a review. American journal of epidemiology 1983;118(2):166-91.

[6] Gordon MO, Beiser JA, Brandt JD, et al. The Ocular Hypertension Treatment Study: Baseline factors that predict the onset of primary open-angle glaucoma. Archives of Ophthalmology 2002;120(6): 714-20.

[7] Kass MA, Heuer DK, Higginbotham EJ et al. The Ocular Hypertension Treatment Study: a randomized trial determines that topical ocular hypotensive medication delays or prevents the onset of primary open-angle glaucoma. Archives of ophthalmology 2002;120(6):701-13.

[8] Foster PJ, Buhrmann R, Quigley HA, et al. The definition and classification of glaucoma in prevalence surveys. British journal of ophthalmology 2002;86(2):238-42.

[9] Chamard C, Villain M, Bron A, et al. Prevalence of unknown ocular hypertension, glaucoma suspects, and glaucoma in patients seen in an ophthalmology center in France. Ophthalmic research 2020;63(3): 295-301.

[10] Varma R, Ying-Lai M, Francis BA, et al. Prevalence of open-angle glaucoma and ocular hypertension in Latinos: the Los Angeles Latino Eye Study. Ophthalmology 2004;111(8):1439-48.

[11] Mitchell P, Smith W, Attebo K, et al. Prevalence of Open-angle Glaucoma in Australia: The Blue Mountains Eye Study. Ophthalmology [Internet] 1996;103(10):1661-9.

[12] Bonomi L, Marchini G, Marraffa M, et al. Prevalence of glaucoma and intraocular pressure distribution in a defined population: the Egna-Neumarkt Study. Ophthalmology 1998;105(2):209-15.

[13] Leske MC, Connell AMS, Wu S-Y, et al. Distribution of intraocular pressure: the Barbados Eye Study. Archives of ophthalmology 1997;115(8):1051-7.

[14] Jacob A, Thomas R, Koshi SP, et al. Prevalence of primary glaucoma in an urban south Indian population. Indian journal of ophthalmology 1998;46(2):81.

[15] Dandona L, Dandona R, Srinivas M, et al. Open-angle glaucoma in an urban population in southern India: the Andhra Pradesh eye disease study. Ophthalmology 2000;107(9):1702-9.

[16] Georgopoulos G, Andreanos D, Liokis N, et al. Risk factors in ocular hypertension. European journal of ophthalmology 1997;7(4):357-63.

[17] Gordon MO, Kass MA. What We Have Learned From the Ocular

Hypertension Treatment Study. American Journal of Ophthalmology 2018;189:xxiv–xxvii.

[18] Thomas R, George R, Parikh R, et al. Five year risk of progression of primary angle closure suspects to primary angle closure: a population based study. British Journal of Ophthalmology 2003;87(4):450 LP – 454.

[19] Weinreb RN, Friedman DS, Fechtner RD, et al. Risk assessment in the management of patients with ocular hypertension. American Journal of Ophthalmology 2004;138(3):458-67.

[20] Quaranta L, Riva I, Gerardi C, et al. Quality of life in glaucoma: a review of the literature. Advances in therapy 2016;33(6):959-81.

[21] Stewart WC, Stewart JA, Nassar QJ, et al. Cost-effectiveness of Treating Ocular Hypertension. Ophthalmology 2008;115(1):94-8.

[22] Wolfram C, Lorenz K, Breitscheidel L, et al. Health-and vision-related quality of life in patients with ocular hypertension or primary open-angle glaucoma. Ophthalmologica 2013;229(4):227-34.

[23] Kazaki M, Georgalas I, Damanakis A, et al. Vision-related quality of life in ocular hypertension patients: Effects of treatment. Open Journal of Ophthalmology 2015;5(01):31.

[24] Kymes SM, Kass MA, Anderson DR, et al. Management of Ocular Hypertension: A Cost-effectiveness Approach From the Ocular Hypertension Treatment Study. American Journal of Ophthalmology 2006;141(6):997-1008.e3.

[25] Quigley HA, Addicks EM, Green WR, et al. Optic nerve damage in human glaucoma: II. The site of injury and susceptibility to damage. Archives of ophthalmology 1981;99(4):635-49.

[26] Burgoyne CF, Downs JC, Bellezza AJ, et al. The optic nerve head as a biomechanical structure: a new paradigm for understanding the role of IOP-related stress and strain in the pathophysiology of glaucomatous optic nerve head damage. Progress in retinal and eye research 2005;24(1):39-73.

[27] Fechtner RD, Weinreb RN. Mechanisms of optic nerve damage in primary open angle glaucoma. Survey of ophthalmology 1994;39(1):23-42.

[28] Yücel YH, Zhang Q, Weinreb RN, et al. Atrophy of relay neurons in magno-and parvocellular layers in the lateral geniculate nucleus in experimental glaucoma. Investigative ophthalmology & visual science 2001;42(13):3216-22.

[29] Weinreb RN, Aung T, Medeiros FA. The pathophysiology and treatment of glaucoma: a review. Jama 2014;311(18):1901-11.

[30] Morgan WH, Yu D-Y, Cooper RL, et al. The influence of cerebrospinal fluid pressure on the lamina cribrosa tissue pressure gradient. Investigative ophthalmology & visual science 1995;36(6):1163-72.

[31] Goel M, Picciani RG, Lee RK, et al. Aqueous humor dynamics: a review. The open ophthalmology journal 2010;4:52.

[32] Tamm ER, Braunger BM, Fuchshofer R. Intraocular pressure and the mechanisms involved in resistance of the aqueous humor flow in the trabecular meshwork outflow pathways. Progress in molecular biology and translational science 2015;134:301-14.

[33] Wang YX, Xu L, Wei WB, et al. Intraocular pressure and its normal range adjusted for ocular and systemic

parameters. The Beijing Eye Study 2011. PLoS One 2018;13(5):e0196926.

[34] Doshi A, Kreidl KO, Lombardi L, et al. Nonprogressive glaucomatous cupping and visual field abnormalities in young Chinese males. Ophthalmology 2007;114(3):472-9.

[35] Casson RJ, Chidlow G, Wood JPM, et al. Definition of glaucoma: clinical and experimental concepts. Clinical & experimental ophthalmology 2012;40(4):341-9.

[36] Mark LP, Prost RW, Ulmer JL, et al. Pictorial review of glutamate excitotoxicity: fundamental concepts for neuroimaging. American journal of neuroradiology 2001;22(10):1813-24.

[37] Neal MJ, Cunningham JR, Hutson PH, et al. Effects of ischaemia on neurotransmitter release from the isolated retina. Journal of neurochemistry 1994;62(3):1025-33.

[38] Liu B, Neufeld AH. Nitric oxide synthase-2 in human optic nerve head astrocytes induced by elevated pressure in vitro. Archives of Ophthalmology 2001;119(2):240-5.

[39] Siu AW, Leung MCP, To CH, et al. Total retinal nitric oxide production is increased in intraocular pressure-elevated rats. Experimental eye research 2002;75(4):401-6.

[40] Liu J, Zhao M-L, Brosnan CF, et al. Expression of type II nitric oxide synthase in primary human astrocytes and microglia: role of IL-1beta and IL-1 receptor antagonist. The Journal of Immunology 1996;157(8):3569-76.

[41] Nathanson JA, McKee M. Alterations of ocular nitric oxide synthase in human glaucoma. Investigative ophthalmology & visual science 1995;36(9):1774-84.

[42] Liu B, Neufeld AH. Expression of nitric oxide synthase-2 (NOS-2) in reactive astrocytes of the human glaucomatous optic nerve head. Glia 2000;30(2):178-86.

[43] Kaur C, Foulds WS, Ling E-A. Hypoxia-ischemia and retinal ganglion cell damage. Clinical Ophthalmology (Auckland, NZ) 2008;2(4):879.

[44] Tezel G, Yang X, Luo C, et al. Hemoglobin expression and regulation in glaucoma: insights into retinal ganglion cell oxygenation. Investigative ophthalmology & visual science 2010;51(2):907-19.

[45] Zheng Y, Wong TY, Mitchell P, et al. Distribution of Ocular Perfusion Pressure and Its Relationship with Open-Angle Glaucoma: The Singapore Malay Eye Study. Investigative Ophthalmology & Visual Science 2010;51(7):3399-404.

[46] Costa VP, Harris A, Anderson D, et al. Ocular perfusion pressure in glaucoma. Acta Ophthalmologica 2014;92(4):e252-66.

[47] Cherecheanu AP, Garhofer G, Schmidl D, et al. Ocular perfusion pressure and ocular blood flow in glaucoma. Current Opinion in Pharmacology 2013;13(1):36-42.

[48] Pasquale LR. Vascular and autonomic dysregulation in primary open-angle glaucoma. Current opinion in ophthalmology 2016;27(2):94.

[49] Stroman GA, Stewart WC, Golnik KC, et al. Magnetic resonance imaging in patients with low-tension glaucoma. Archives of ophthalmology 1995;113(2):168-72.

[50] Ong K, Farinelli A, Billson F, et al. Comparative study of brain magnetic resonance imaging findings in patients with low-tension glaucoma and control subjects. Ophthalmology 1995;102(11): 1632-8.

[51] Neufeld AH, Kawai S, Das S, et al. Loss of retinal ganglion cells following retinal ischemia: the role of inducible nitric oxide synthase. Experimental eye research 2002;75(5):521-8.

[52] Wu S-Y, Leske MC. Associations with intraocular pressure in the Barbados Eye Study. Archives of ophthalmology 1997;115(12):1572-6.

[53] Kim E, Varma R. Glaucoma in Latinos/Hispanics. Current Opinion in Ophthalmology 2010;21(2).

[54] EGPS Group. Miglior S, Pfeiffer N, Torri V et al. Predictive Factors for Open-Angle Glaucoma among Patients with Ocular Hypertension in the European Glaucoma Prevention Study. Ophthalmology 2007;114(1):3-9.

[55] Landers J, Goldberg I, Graham SL. Analysis of risk factors that may be associated with progression from ocular hypertension to primary open angle glaucoma. Clinical and Experimental Ophthalmology 2002;30(4):242-7.

[56] Bengtsson B, Heijl A. A long-term prospective study of risk factors for glaucomatous visual field loss in patients with ocular hypertension. Journal of glaucoma 2005;14(2):135-8.

[57] The AGIS Investigators. The Advanced Glaucoma Intervention Study (AGIS): 7. The relationship between control of intraocular pressure and visual field deterioration. The AGIS Investigators. Am J Ophthalmol 2000;130(4):429-40.

[58] Le A, Mukesh BN, McCarty CA, et al. Risk factors associated with the incidence of open-angle glaucoma: the visual impairment project. Investigative ophthalmology & visual science 2003;44(9):3783-9.

[59] Leske MC, Heijl A, Hussein M, et al. Factors for glaucoma progression and the effect of treatment: the early manifest glaucoma trial. Archives of ophthalmology 2003;121(1):48-56.

[60] Wensor MD, McCarty CA, Stanislavsky YL, et al. The prevalence of glaucoma in the Melbourne Visual Impairment Project. Ophthalmology 1998;105(4):733-9.

[61] Collaborative Normal-Tension Glaucoma Study Group. Comparison of glaucomatous progression between untreated patients with normal-tension glaucoma and patients with therapeutically reduced intraocular pressures. American journal of ophthalmology 1998;126(4):487-97.

[62] Heijl A, Leske MC, Bengtsson B, et al. Reduction of intraocular pressure and glaucoma progression: results from the Early Manifest Glaucoma Trial. Archives of ophthalmology 2002;120(10):1268-79.

[63] Konstas AGP, Irkec MT, Teus MA, et al. Mean intraocular pressure and progression based on corneal thickness in patients with ocular hypertension. Eye 2009;23(1):73-8.

[64] Damji KF, Muni RH, Munger RM. Influence of corneal variables on accuracy of intraocular pressure measurement. Journal of glaucoma 2003;12(1):69-80.

[65] Liu J, Roberts CJ. Influence of corneal biomechanical properties on intraocular pressure measurement: Quantitative analysis. Journal of Cataract & Refractive Surgery 2005;31(1):146-55.

[66] Zakrzewska A, Wiącek MP, Machalińska A. Impact of corneal parameters on intraocular pressure measurements in different tonometry methods. International journal of ophthalmology 2019;12(12):1853-8.

[67] Argus WA. Ocular hypertension and central corneal thickness. Ophthalmology 1995;102(12):1810-2.

[68] Herndon LW, Choudhri SA, Cox T, et al. Central corneal thickness in normal, glaucomatous, and ocular hypertensive eyes. Archives of Ophthalmology 1997;115(9):1137-41.

[69] Joda AA, Shervin MMS, Kook D, et al. Development and validation of a correction equation for Corvis tonometry. Computer methods in biomechanics and biomedical engineering 2016;19(9):943-53.

[70] Eliasy A, Chen K-J, Vinciguerra R, et al. Ex-vivo experimental validation of biomechanically-corrected intraocular pressure measurements on human eyes using the CorVis ST. Experimental eye research 2018;175:98-102.

[71] Nemesure B, Wu S-Y, Hennis A, Group BES. Corneal thickness and intraocular pressure in the Barbados eye studies. Archives of Ophthalmology 2003;121(2):240-4.

[72] la Rosa FA, Gross RL, Orengo-Nania S. Central corneal thickness of Caucasians and African Americans in glaucomatous and nonglaucomatous populations. Archives of ophthalmology 2001;119(1):23-7.

[73] Friedman DS, Wilson MR, Liebmann JM, et al. An evidence-based assessment of risk factors for the progression of ocular hypertension and glaucoma. American Journal of Ophthalmology 2004;138(3 SUPPL.):19-31.

[74] Budenz DL, Huecker JB, Gedde SJ, et al. Group OHTS. Thirteen-Year Follow-up of Optic Disc Hemorrhages in the Ocular Hypertension Treatment Study. American journal of ophthalmology 2017;174:126-33.

[75] Daubs JG, Crick RP. Effect of refractive error on the risk of ocular hypertension and open angle glaucoma. Transactions of the ophthalmological societies of the United Kingdom 1981;101(1):121-6.

[76] Perkins ES, Phelps CD. Open angle glaucoma, ocular hypertension, low-tension glaucoma, and refraction. Archives of Ophthalmology 1982;100(9):1464-7.

[77] Ponte F, GiuffrÉ G, Giammanco R, Risk factors of ocular hypertension and glaucoma - The Casteldaccia Eye Study. Documenta Ophthalmologica 1994;85(3):203-10.

[78] Budenz DL, Anderson DR, Feuer WJ, et al. Detection and Prognostic Significance of Optic Disc Hemorrhages during the Ocular Hypertension Treatment Study. Ophthalmology 2006;113(12):2137-43.

[79] Miglior S, Torri V, Zeyen T, et al. Intercurrent factors associated with the development of open-angle glaucoma in the European glaucoma prevention study. American journal of ophthalmology 2007;144(2):266-75.

[80] Heijl A, Åsman P. A clinical study of perimetric probability maps. Archives of Ophthalmology 1989;107(2):199-203.

[81] Bengtsson B, Heijl A. A visual field index for calculation of glaucoma rate of progression. American journal of ophthalmology 2008;145(2):343-53.

[82] Mills RP, Budenz DL, Lee PP, et al. Categorizing the stage of glaucoma from pre-diagnosis to end-stage disease. American journal of ophthalmology 2006;141(1):24-30.

[83] Åsman P, Heijl A. Glaucoma hemifield test: automated visual field evaluation. Archives of ophthalmology 1992;110(6):812-9.

[84] Heijl A, Bengtsson B, Chauhan BC, et al. A comparison of visual field progression criteria of 3 major glaucoma trials in early manifest glaucoma trial patients. Ophthalmology 2008;115(9):1557-65.

[85] Sample PA, Bosworth CF, Blumenthal EZ, et al. Visual function–specific perimetry for indirect comparison of different ganglion cell populations in glaucoma. Investigative ophthalmology & visual science 2000;41(7):1783-90.

[86] Sample PA. Short-wavelength automated perimetry: it's role in the clinic and for understanding ganglion cell function. Progress in retinal and eye research 2000;19(4):369-83.

[87] Polo V, Larrosa JM, Pinilla I, et al. Predictive value of short-wavelength automated perimetry: a 3-year follow-up study. Ophthalmology 2002;109(4):761-5.

[88] Landers J, Goldberg I, Graham S. A comparison of short wavelength automated perimetry with frequency doubling perimetry for the early detection of visual field loss in ocular hypertension. Clinical & experimental ophthalmology 2000;28(4):248-52.

[89] Liu S, Lam S, Weinreb RN, et al. Comparison of standard automated perimetry, frequency-doubling technology perimetry, and short-wavelength automated perimetry for detection of glaucoma. Investigative ophthalmology & visual science 2011;52(10):7325-31.

[90] Klewin KM, Radius RL. Background illumination and automated perimetry. Archives of Ophthalmology 1986;104(3):395-7.

[91] Heuer DK, Anderson DR, Feuer WJ, et al. The influence of decreased retinal illumination on automated perimetric threshold measurements. American journal of ophthalmology 1989;108(6):643-50.

[92] Henson DB, Morris EJ. Effect of uncorrected refractive errors upon central visual field testing. Ophthalmic and Physiological Optics 1993;13(4):339-43.

[93] Haas A, Flammer J, Schneider U. Influence of age on the visual fields of normal subjects. American journal of ophthalmology 1986;101(2):199-203.

[94] Lindenmuth KA, Skuta GL, Rabbani R, et al. Effects of pupillary dilation on automated perimetry in normal patients. Ophthalmology 1990;97(3):367-70.

[95] Lindenmuth KA, Skuta GL, Rabbani R, et al. Effects of pupiliary constriction on automated perimetry in normal eyes. Ophthalmology 1989;96(9):1298-301.

[96] Demirel S, de Moraes CG v, Gardiner SK, et al. The rate of visual field change in the ocular hypertension treatment study. Investigative ophthalmology & visual science 2012;53(1):224-7.

[97] Soliman MAE, de Jong LAMS, Ismaeil A-AA, et al. Standard achromatic perimetry, short wavelength automated perimetry, and frequency doubling technology for detection of glaucoma damage. Ophthalmology 2002;109(3):444-54.

[98] Johnson CA, Brandt JD, Khong AM, et al. Short-wavelength automated perimetry in low-, medium-, and high-risk ocular hypertensive eyes: initial baseline results. Archives of Ophthalmology 1995;113(1):70-6.

[99] Bengtsson B, Heijl A. Diagnostic sensitivity of fast blue–yellow and standard automated perimetry in early glaucoma: a comparison between different test programs. Ophthalmology 2006;113(7):1092-7.

[100] de Moraes CG, Demirel S, Gardiner SK, et al. Effect of treatment on the rate of visual field change in the ocular hypertension treatment study

observation group. Investigative ophthalmology & visual science 2012;53(4):1704-9.

[101] Chow C, Chan PP. Risk Assessment of Ocular Hypertension and the Use of Medication. In: Visual Impairment and Blindness-What We Know and What We Have to Know. IntechOpen; 2020.

[102] Musch DC, Gillespie BW, Lichter PR, et al. Visual field progression in the Collaborative Initial Glaucoma Treatment Study: the impact of treatment and other baseline factors. Ophthalmology 2009;116(2):200-7.

[103] Miglior S, Zeyen T, Pfeiffer N, et al. Results of the European Glaucoma Prevention Study. Ophthalmology 2005;112(3):366-75.

[104] Medeiros FA, Weinreb RN, Sample PA, et al. Validation of a predictive model to estimate the risk of conversion from ocular hypertension to glaucoma. Archives of Ophthalmology 2005;123(10):1351-60.

[105] Drance S, Anderson DR, Schulzer M, Group CN-TGS. Risk factors for progression of visual field abnormalities in normal-tension glaucoma. American journal of ophthalmology 2001;131(6):699-708.

[106] Brandt JD. Corneal thickness in glaucoma screening, diagnosis, and management. Current opinion in ophthalmology 2004;15(2):85-9.

[107] Medeiros FA, Sample PA, Zangwill LM, et al. Corneal thickness as a risk factor for visual field loss in patients with preperimetric glaucomatous optic neuropathy. American journal of ophthalmology 2003;136(5):805-13.

[108] Fan X, Wu L-L, Ma Z-Z, et al. Usefulness of frequency-doubling technology for perimetrically normal eyes of open-angle glaucoma patients with unilateral field loss. Ophthalmology 2010;117(8):1530-7.

[109] Cao KY, Kapasi M, Betchkal JA, et al. Relationship between central corneal thickness and progression of visual field loss in patients with open-angle glaucoma. Canadian Journal of Ophthalmology 2012;47(2):155-8.

[110] Flammer J, Orgül S, Costa VP, et al. The impact of ocular blood flow in glaucoma. Progress in retinal and eye research 2002;21(4):359-93.

[111] Pache M, Flammer J. A sick eye in a sick body? Systemic findings in patients with primary open-angle glaucoma. Survey of ophthalmology 2006;51(3):179-212.

[112] Grødum K, Heijl A, Bengtsson B. Risk of glaucoma in ocular hypertension with and without pseudoexfoliation. Ophthalmology 2005;112(3):386-90.

[113] Fitzke FW, Hitchings RA, Poinoosawmy D, et al. Analysis of visual field progression in glaucoma. British Journal of Ophthalmology 1996;80(1):40-8.

[114] Heijl A, Leske MC, Bengtsson B, Group EMGT. Measuring visual field progression in the Early Manifest Glaucoma Trial. Acta Ophthalmologica Scandinavica 2003;81(3):286-93.

[115] Zangwill LM, Weinreb RN, Beiser JA, et al. Baseline topographic optic disc measurements are associated with the development of primary open-angle glaucoma: the Confocal Scanning Laser Ophthalmoscopy Ancillary Study to the Ocular Hypertension Treatment Study. Archives of ophthalmology 2005;123(9):1188-97.

[116] Weinreb RN, Zangwill LM, Jain S, et al. Predicting the onset of glaucoma: the confocal scanning laser ophthalmoscopy ancillary study to the

Ocular Hypertension Treatment Study. Ophthalmology 2010;117(9):1674-83.

[117] Strouthidis NG, Scott A, Peter NM, et al. Optic disc and visual field progression in ocular hypertensive subjects: detection rates, specificity, and agreement. Investigative ophthalmology & visual science 2006;47(7):2904-10.

[118] Leung CKS, Liu S, Weinreb RN, et al. Evaluation of retinal nerve fiber layer progression in glaucoma: a prospective analysis with neuroretinal rim and visual field progression. Ophthalmology 2011;118(8):1551-7.

[119] Fayers T, Strouthidis NG, Garway-Heath DF. Monitoring glaucomatous progression using a novel Heidelberg Retina Tomograph event analysis. Ophthalmology 2007;114(11):1973-80.

[120] Strouthidis NG, Demirel S, Asaoka R, et al. The Heidelberg Retina Tomograph Glaucoma Probability Score: Reproducibility and Measurement of Progression. Ophthalmology 2010;117(4):724-9.

[121] Alencar LM, Bowd C, Weinreb RN, et al. Comparison of HRT-3 glaucoma probability score and subjective stereophotograph assessment for prediction of progression in glaucoma. Investigative ophthalmology & visual science 2008;49(5):1898-906.

[122] Pederson JE, Anderson DR. The mode of progressive disc cupping in ocular hypertension and glaucoma. Archives of ophthalmology 1980;98(3):490-5.

[123] Tuulonen A, Airaksinen PJ. Initial glaucomatous optic disk and retinal nerve fiber layer abnormalities and their progression. American journal of ophthalmology 1991;111(4):485-90.

[124] Odberg T, Riise D. Early diagnosis of glaucoma: the value of successive stereophotography of the optic disc. Acta ophthalmologica 1985;63(3):257-63.

[125] Lloyd MJ, Mansberger SL, Fortune BA, et al. Features of optic disc progression in patients with ocular hypertension and early glaucoma. Journal of glaucoma 2013;22(5):343.

[126] Tezel G, Kolker AE, Wax MB, et al. Parapapillary chorioretinal atrophy in patients with ocular hypertension: II. An evaluation of progressive changes. Archives of ophthalmology 1997;115(12):1509-14.

[127] Tezel G, Kolker AE, Kass MA, et al. Parapapillary chorioretinal atrophy in patients with ocular hypertension: I. An evaluation as a predictive factor for the development of glaucomatous damage. Archives of ophthalmology 1997;115(12):1503-8.

[128] Kubota T, Jonas JB, Naumann GO. Direct clinico-histological correlation of parapapillary chorioretinal atrophy. British journal of ophthalmology 1993;77(2):103-6.

[129] Jonas JB, Naumann GO. Parapapillary chorioretinal atrophy in normal and glaucoma eyes. II. Correlations. Investigative ophthalmology & visual science 1989;30(5):919-26.

[130] Jonas JB, Fernández MC, Stürmer J. Pattern of glaucomatous neuroretinal rim loss. Ophthalmology 1993;100(1):63-8.

[131] Airaksinen PJ, Tuulonen A, Alanko HI. Rate and Pattern of Neuroretinal Rim Area Decrease in Ocular Hypertension and Glaucoma. Archives of Ophthalmology 1992;110(2):206-10.

[132] Tielsch JM, Katz J, Quigley HA, et al. Intraobserver and Interobserver Agreement in Measurement of Optic

Disc Characteristics. Ophthalmology 1988;95(3):350-6.

[133] Yang B, Ye C, Yu M, et al. Optic disc imaging with spectral-domain optical coherence tomography: variability and agreement study with Heidelberg retinal tomograph. Ophthalmology 2012;119(9):1852-7.

[134] Reis ASC, Burgoyne CF, Chauhan BC. Optic Disc Margin Anatomy in Glaucoma Patients and Normal Controls with SD-OCT.pdf. Ophthalmology 2012;119(4):738-47.

[135] Chauhan BC, O'Leary N, AlMobarak FA, et al. Enhanced detection of open-angle glaucoma with an anatomically accurate optical coherence tomography-derived neuroretinal rim parameter. Ophthalmology 2013;120(3):535-43.

[136] Chan PP, Chiu V, Wong MO. Variability of vertical cup to disc ratio measurement and the effects of glaucoma 5-year risk estimation in untreated ocular hypertensive eyes. British Journal of Ophthalmology 2019;103(3):361 – 368.

[137] Zheng F, Wu Z, Leung CKS. Detection of Bruch's Membrane Opening in Healthy Individuals and Glaucoma Patients with and without High Myopia. Ophthalmology 2018;125(10):1537-46.

[138] Sawada Y, Araie M, Shibata H, et al. Optic Disc Margin Anatomic Features in Myopic Eyes with Glaucoma with Spectral-Domain OCT. Ophthalmology 2018;125(12):1886-97.

[139] Jonas JB, Wang YX, Dong L, et al. Advances in myopia research anatomical findings in highly myopic eyes. Eye and Vision 2020;7(1):1-10.

[140] Kass MA, Gordon MO, Gao F, et al. Delaying treatment of ocular hypertension: The ocular hypertension treatment study. Archives of Ophthalmology 2010;128(3):276-87.

[141] Medeiros FA, Weinreb RN, Zangwill LM, et al. Long-term intraocular pressure fluctuations and risk of conversion from ocular hypertension to glaucoma. Ophthalmology 2008;115(6):934-40.

[142] Bengtsson B, Heijl A. Diurnal IOP fluctuation: not an independent risk factor for glaucomatous visual field loss in high-risk ocular hypertension. Graefe's Archive for Clinical and Experimental Ophthalmology 2005;243(6):513-8.

[143] Prum BE, Lim MC, Mansberger SL, et al. Primary open-angle glaucoma suspect preferred practice pattern® guidelines. Ophthalmology 2016;123(1):P112-51.

[144] Kaufman PL. Latanoprostene bunod ophthalmic solution 0.024% for IOP lowering in glaucoma and ocular hypertension. Expert Opinion on Pharmacotherapy 2017;18(4):433-44.

[145] Vittitow JL, Liebmann JM, Kaufman PL, et al. Long-term efficacy and safety of latanoprostene bunod 0.024% for intraocular pressure lowering in patients with open-angle glaucoma or ocular hypertension: APOLLO and LUNAR studies. Investigative Ophthalmology & Visual Science 2016;57(12):3030.

[146] Kawase K, Vittitow JL, Weinreb RN et al. Long-term safety and efficacy of latanoprostene bunod 0.024% in Japanese subjects with open-angle glaucoma or ocular hypertension: the JUPITER study. Advances in therapy 2016;33(9):1612-27.

[147] Gazzard G, Konstantakopoulou E, Garway-Heath D, et al. Selective laser trabeculoplasty versus eye drops for first-line treatment of ocular hypertension and glaucoma (LiGHT): a

multicentre randomised controlled trial. Lancet (London, England) 2019;393(10180):1505-16.

[148] Gazzard G, Konstantakopoulou E, Garway-Heath D, et al. Laser in Glaucoma and Ocular Hypertension (LiGHT) trial. A multicentre, randomised controlled trial: design and methodology. British Journal of Ophthalmology 2018;102(5):593 – 598.

[149] Wright DM, Konstantakopoulou E, Montesano G, et al. Visual Field Outcomes from the Multicenter, Randomized Controlled Laser in Glaucoma and Ocular Hypertension Trial (LiGHT). Ophthalmology 2020;127(10):1313-21.

[150] Garg A, Nathwani N, Garway-Heath D, et al. Primary Selective Laser Trabeculoplasty for Open-Angle Glaucoma and Ocular Hypertension: Clinical Outcomes, Predictors of Success, and Safety from the Laser in Glaucoma and Ocular Hypertension Trial. Ophthalmology 2019;126(9):1238-48.

Chapter 4

Neuropathology in Hypertensive Glaucoma

Jan Lestak and Martin Fůs

Abstract

Hypertensive glaucoma is still defined as a disease where, at high intraocular pressure, retinal ganglion cell axons are impaired with excavation at the optic disc and changes in the visual field. In single cases, the study highlights the importance of knowledge of neuropathology not only at the level of the retina but the entire visual pathway, including the visual centres in the brain. It uses the issue of neurotransmission in the visual analyser and its pathology, but mainly the results of electrophysiological examinations and functional imaging of the brain using Positron Emission Tomography and Functional Magnetic Resonance. It does not overlook the imaging methods of the eye (nerve fibre layer, vessel density). On the basis of this information, therapy is recommended as well.

Keywords: physiology and pathology of neurotransmission in the visual pathway, hypertensive glaucoma, neuropathology, electrophysiology, fMRI, PET, therapy

1. Introduction

For correct and early diagnosis of glaucoma, it is useful to know the processes that occur in the eye during pathological intraocular pressure (IOP). In a chronic condition, when the nerve structures are no longer able to cope with this cause, the process also expands vertically. Subsequently, even following compensation of IOP, the disease progresses. In this study, you will read about the processes explaining the pathology in the entire visual pathway and the treatment options consisting not only of the reduction of IOP.

2. Electrophysiology in the visual analyser

One of the first stimuli that led us to examine glaucomas was the simultaneous measurement of pattern electroretinogram (PERG) and pattern visual evoked potentials (PVEP) in a 20-year-old healthy individual –firstly at IOP of 15 mmHg and then after increasing it to 40 mmHg. To our surprise, neurotransmission was blocked at the retinal ganglion cell level, while PVEPs changed slightly (**Figure 1**). With the blockade of transmission at the level of ganglion cells, we also expected an unequivocal or at least abnormal PVEP response (measurements were performed in 1988). This fact did not fit into the existing definitions of glaucoma regarding retinal ganglion cell axon dysfunction with excavation at the optic disc and changes in the visual field [1, 2].

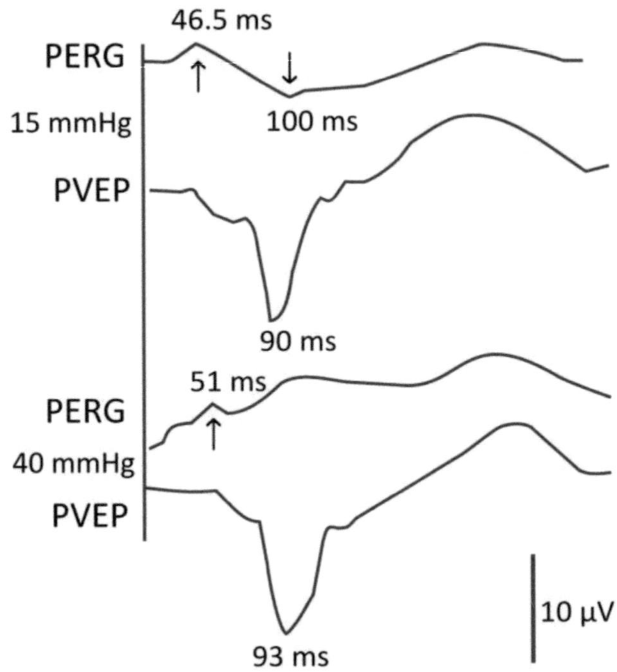

Figure 1.
Upper curve (PERG) and below it PVEP at normal IOP. The arrows indicate the latencies of oscillations, the amplitude of which arises from the responses of retinal ganglion cells. The lower curves - following increase of IOP to 40 mmHg [3].

Therefore, we tried to find an answer to this reaction of the visual analyser. Several questions remained unanswered. Why did the retinal ganglion cells not respond? What happened to the central visual pathway when, following the blockade at the retinal ganglion cell level, we elicited an almost "normal response" from the brain? To an electrophysiologist, there was only one explanation. After stabilisation of binocular functions, the visual cortex is set to receive a certain amount of action potentials. When it is reduced at any level, from photoreceptors to cortical cells of the brain, it begins to stimulate the visual pathway through feedback processes in order to restore their numbers to their original values [4–7].

Before elucidating the above mechanisms, we briefly explain the process of transmitting electrical changes in the visual pathway, from photoreceptors to the visual centres in the brain.

3. Neurotransmission in the visual analyser

After the impact of light on the retina, the cis-retinal chemically changes its outer form in the outer segments of photoreceptors, leading to their hyperpolarisation [8].

Hyperpolarisation of photoreceptors causes the release of glutamate from the presynaptic neuron into the synaptic cleft during synaptic transmission and subsequent binding to receptors located on the membrane of the postsynaptic neuron [9]. Glutamate binds to receptors that have been named after their selective agonists. A typical agonist for NMDA receptors is N-methyl-D-aspartate. For AMPA receptors, α-amino-3-hydroxy-5-methyl-4-8 isoxazole propionate (AMPA), and for the third type, kainate receptors, kainate. AMPA and kainate receptors are also referred to as non-NMDA. NMDA receptors are ion channels permeable to calcium (Ca) ions. The flow of Ca through the NMDA receptors is blocked by magnesium

(Mg) ions at normal membrane potential. This block can be eliminated by strong depolarisation [10, 11]. Excessive flow of Ca into the cell through NMDA voltage-gated channels can cause hypoxia, hypoglycaemia, etc. Under these conditions, glutamate levels in the synaptic cleft remain elevated for a long time, with NMDA receptors being permanently activated, leading to intracellular Ca concentrations that are cytotoxic [12]. The concentration of free glutamate in the synaptic cleft reaches approximately 1.1 mM during the synaptic transmission. However, its concentration decreases rapidly and breaks down at NMDA receptors within 1.2 ms. However, glutamate dissociates much faster from the AMPA receptors. Thus, the time course of free glutamate predicts that dissociation contributes to AMPA receptor-mediated postsynaptic current breakdown. Otherwise, voltage-gated channels would open [9].

In the mammalian central nervous system, glutamate is primarily removed from the synapse by the excitatory amino acid transporter (EAAT) glutamate transporters and the glutamate aspartate transporter (GLAST as a glutamate transporter to Muller cells (MB) and glutamine synthetase (GL) as an enzyme that converts glutamate to glutamine in MB) [13, 14]. In the glial cells (MB), glutamate is subsequently converted to glutamine, which no longer acts as a neurotransmitter and can thus be released back into the synapse, from where it is subsequently taken up by a presynaptic neuron, which converts it back to glutamate [15]. To date, there is no evidence of the presence of an enzyme that converts glutamate directly in the synapse [16].

The glutamate receptors are expressed not only in retinal ganglion cells but also in the photoreceptors, as well as in the horizontal and bipolar cells [17]. Glutamate is the predominant excitatory neurotransmitter, not only in the retina, but also in the mammalian brain [18].

In this way, the processes are clarified that function in the transmission of electrical changes in voltage in the visual path.

4. Efforts to restore the amount of action potentials entering the brain

We have two options for restoring the amount of action potentials entering the brain to their original values. The first is to flush out a larger amount of neurotransmitter at the level of the "damaged" cell, and the second is to leave this neurotransmitter in the synaptic cleft for a longer period of time. Both variants have been experimentally demonstrated in glaucoma.

In the vitreous body of the glaucoma eyes of experimental animals, the value of glutamate was up to three times higher in comparison with the control group. These values are toxic to both the ganglion cell layer and the inner plexiform layer. The GLAST and GS values did not increase until 3 weeks after an increase in IOP in rats. The number of ganglion cells decreased from 4 to 60 days from an increase in IOP to 6 to 44% [19, 20].

Another important discovery is that the glutamate transporter can begin to function in reverse and transfer glutamate and sodium from the cell back to the synaptic cleft. Thus, the flushed glutamate comes only in a small part from the synaptic vesicles; most of it comes from the cytosol, where it was previously pumped to [21].

The long-term action of glutamate on non-NMDA receptors increases the post-synaptic potential and opens up the voltage-bound receptors which are normally closed by Mg, and the entry of Ca into the cell. This process takes place in all cells with glutamate receptors. Therefore, glaucoma damages not only retinal ganglion cells, but also cells in the inner nuclear layer and the photoreceptor layer [22].

Nerve cells do not die immediately following the entrance of Ca into the cells. As mentioned above, their size will first be reduced. If they have a sufficient energy supply, they will cope with this situation. Once the energy is consumed, an apoptotic or necrotic process begins and the cell dies [23].

With regard to the first of the above questions: why did the retinal ganglion cells not respond? - we found the answer in the study of Morgan et al., Naskar et al. and others who experimentally studied retinal cells after an acute increase in IOP. They found primary changes in the ganglion cells [23–26].

What happened to the central visual pathway when, after blockade at the retinal ganglion cell level, we elicited an almost "normal response" from the brain?

If the visual pathway, including the visual cortex, is also involved in the process of hypertensive glaucoma, then we should also find changes in the brain. Standard structural examination techniques do not allow this diagnosis.

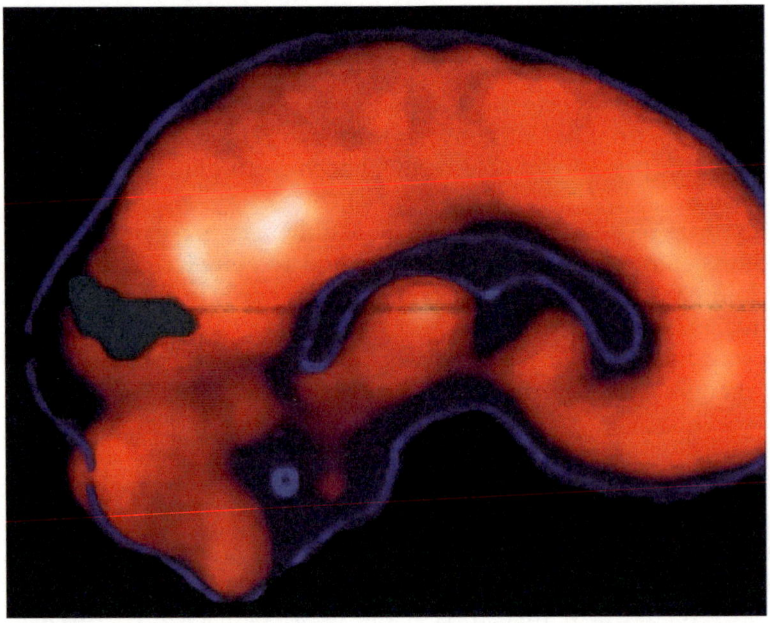

Figure 2.
Sagittal section of the brain of a 58-year-old patient with secondary glaucoma. VOD: 0.05, correct light projection, VOS: 1.0 naturally. c/d = 1.0. The green colour indicates fluorodeoxyglucose deficiency in the visual Centre [27].

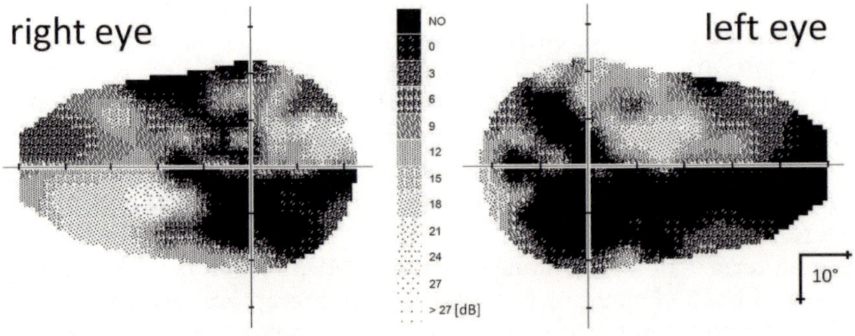

Figure 3.
*Field of view of the same patient as in **Figure 2** from the same period [27].*

For this reason, we sought a solution in Positron Emission Tomography. Radioactive glucose (18-fluororodeoxyglucose) is used to examine brain activity, which is taken up in healthy cells. **Figure 2** shows the absence of glucose radioactivity in the occipital lobe and **Figure 3** shows the field of view of the same patient. We performed the examination in 2001.

Because it was necessary to verify this finding in a larger group of patients with HTG and we did not want to burden the patients with radioactive material, we performed the further examination using Functional Magnetic Resonance Imaging (fMRI) on the visual paradigm (**Figure 4**). As an example, we present the result of the fMRI examination in the same patient, in whom we also performed the PET examination **Figure 5**.

Figure 4.
Chessboard field of black and white stimulation (a) and yellow-blue stimulation (b). During stimulation, the chessboard field alternates with its inversion at a frequency of 2 Hz.

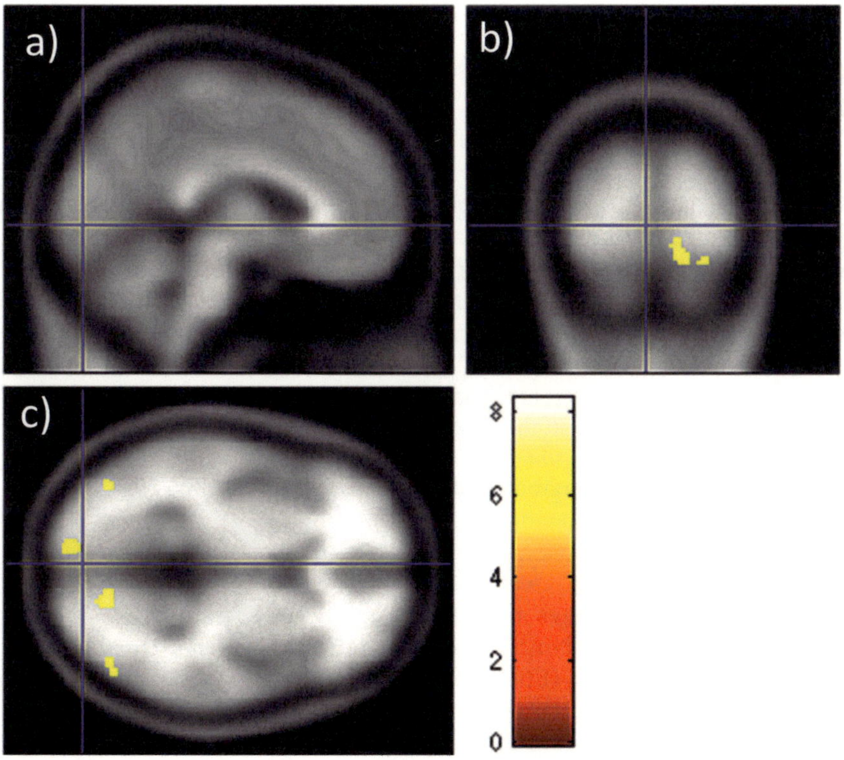

Figure 5.
Functional magnetic resonance imaging (fMRI) following the visual paradigm of a black and white chessboard in the same patient as in ***Figure 2****. We performed the examination in 2010. Sagittal (a), coronary (b) and transverse (c) section [27].*

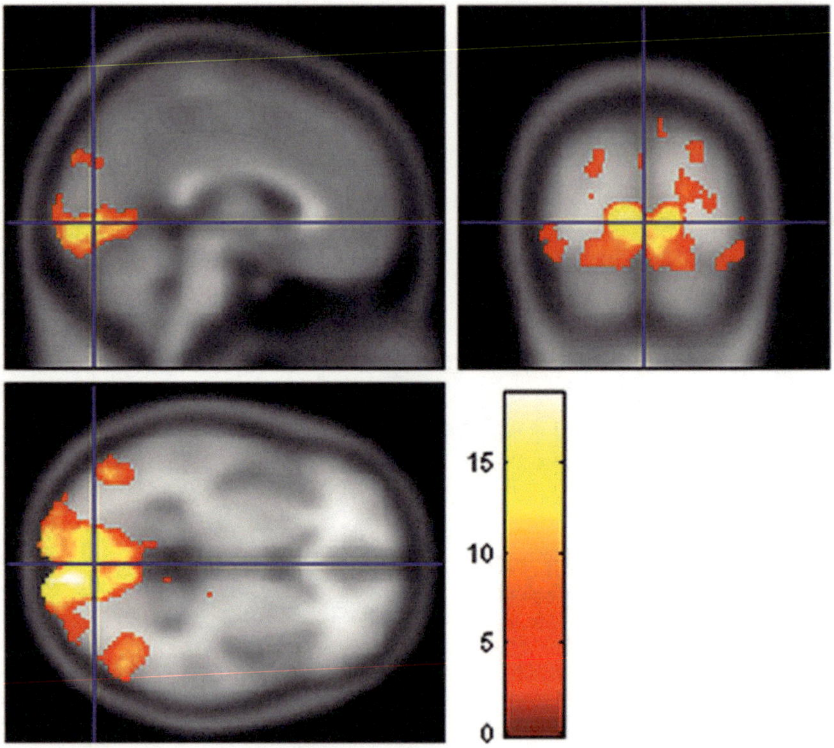

Figure 6.
fMRI in a patient with normotensive glaucoma (61 years), V: 1.0 without correction. c/d = 1 [27].

For comparison, we also present a normal fMRI finding in a patient with normotensive glaucoma (**Figure 6**).

With the use of Positron Emission Tomography and fMRI, we have shown that hypertensive glaucoma also damages the visual centres in the brain. The finding was surprising in patients with HTG following stimulation with both black-white and yellow-blue stimulation, which has not been used in any of the cited work (**Figure 4**). We examined 8 patients with HTG (various stages) and compared their results with the results of 8 healthy people. We found that the difference in the number of activated voxels in patients using black and white versus yellow-blue stimulation was 59%. It was only 2% in a control group. We consider that the ganglion cell damage will be greater than normally expected [27].

5. Electrophysiology in hypertensive glaucoma

We now return to electrophysiological eye examinations.

In experimental glaucoma, electroretinographic changes (decrease in amplitudes of up to 50%) preceded changes in the retinal nerve fibre layer [28].

These facts, as well as the conclusions of other authors, [29–31] led us to use the electrophysiological methods (PERG and PVEP) to determine the level and depth of damage in various types of hypertensive glaucoma. Based on these examinations, we concluded that the changes in PERG and PVEP in HTG indicate that not only the retinal ganglion cells and subsequently their axons, but also the visual centres in the brain are damaged [32].

At the neuronal membrane level, the relationship between the two neurotransmitter systems is supported by the direct inhibition of the NMDA receptor by dopamine and the inhibitory effect of glutamate on dopamine release. This means that a higher dopamine level blocks the NMDA receptor and, conversely, glutamate blocks the release of dopamine [33, 34].

To verify this biochemical information, we used the examination of the oscillating potentials of the electroretinogram. Amacrine cells are divided into dopaminergic and GABAergic according to the neurotransmitter. Dopaminergic cells generate oscillating potentials in the electroretinogram, and GABAergic cells participate in the formation of the threshold scotopic potential [35].

We performed the examination in 2001 on a Primus device (Lacce Elettronica) according to the ISCEV methodology (1989). After a 30-minute adaptation to darkness, we examined the oscillating potentials. Retinal stimulation in artificial mydriasis (tropicamide) was performed with a 5 ms flash light with a luminous flux of 2.5 cd/m^2/s. An average of 10 responses (stimuli after 15 seconds) were averaged, using 80 to 500 Hz filters. We evaluated latency and amplitude of the P2 oscillation.

The first group consisted of 23 eyes of healthy people. In the second group, there were 36 glaucoma eyes with incipient changes in visual field with compensated IOP. Persons included in the groups had an average age of 40.3 years (35–56). The results showed a statistically significant prolongation of the latency of the P2 oscillation ($p = 0.049$) and a reduction of the amplitude of the P2 oscillation ($p = 0.001$) in the glaucoma eyes, compared to the healthy group. In this way, we indirectly demonstrated increased levels of glutamate in glaucoma eyes.

We were also interested in how retinal ganglion cells (PERG) will behave when adjusting antiglaucoma treatment, and subsequently the whole visual analyser (PVEP). We performed PERG and PVEP examinations (according to the ISCEV - 2012 methodology on the Roland Consult Germany device) in a patient (64 years old)

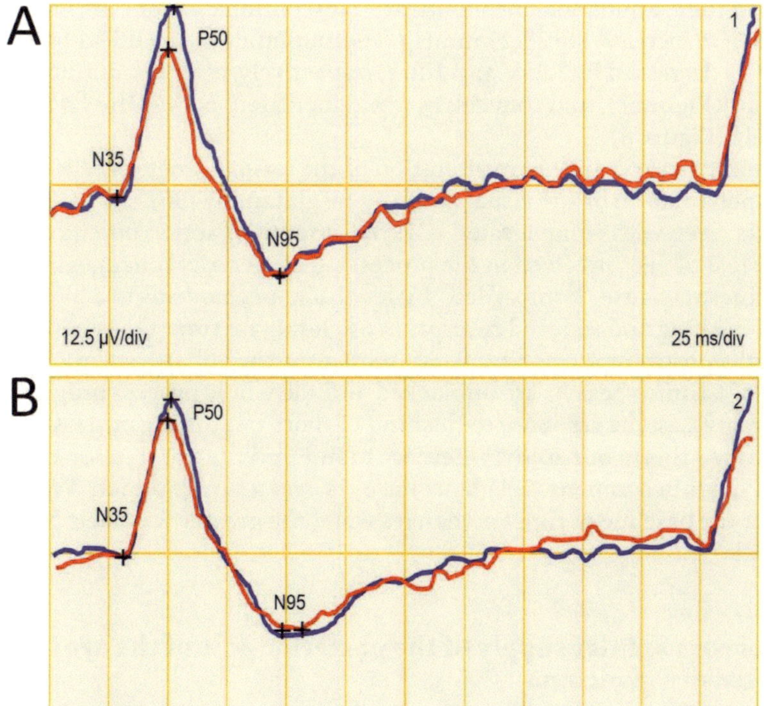

Figure 7.
PERG before discontinuation (blue curve) and after discontinuation of antiglaucoma medications (red curve). Right eye (A), left eye (B).

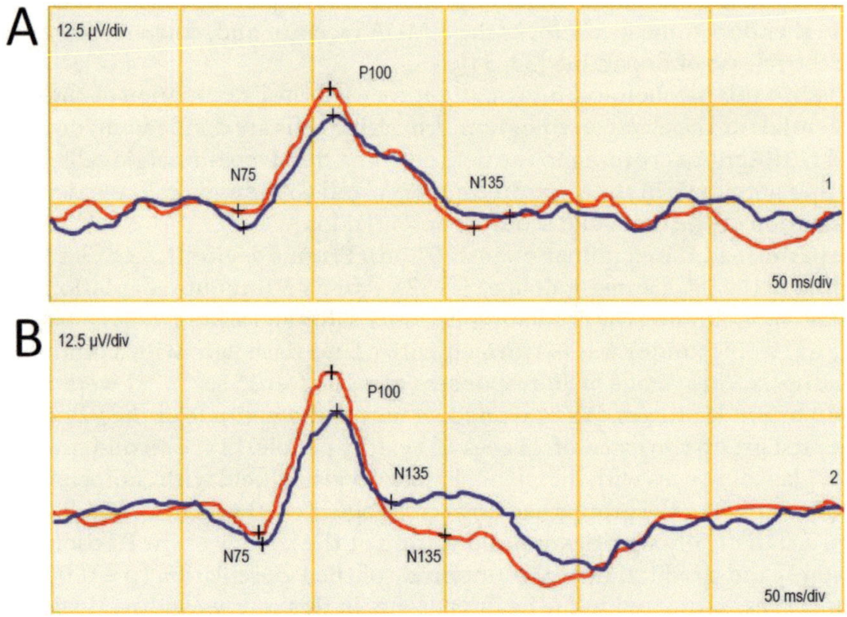

Figure 8.
PVEP before discontinuation (blue curve) and following discontinuation of antiglaucoma medicines (red curve). Following stimulation of the right eye (A) and left eye (B).

with glaucoma, compensated by Cosopt and Lumigan to IOP of 18/18 mmHg. The perimeter was normal in this patient, c/d = 0.4 and GDX 11/20. Due to these values, we repeated the examination one month after discontinuation of both antiglaucoma medicines. IOP increased to 23/29 mmHg. The amplitude of the PERG oscillations P50 and N95 decreased by 3.2uV and 1.1uV, respectively, after discontinuation of the medications (**Figure 7**), and conversely, it was increased in PVEP by 1.4 and 4.7uV, respectively (**Figure 8**).

This finding also indicates an alteration of the retinal ganglion cells and, conversely, a potentiation of the visual pathway by glutamate [36].

Because, even with compensated IOP, the amount of action potentials is reduced due to the loss of cells involved in the processing of electrical changes in the visual pathway, these cells are "bombarded" by feedback mechanisms to a higher response. Excessive flushing and reduced resorption of glutamate from the synaptic cleft increases the postsynaptic potential. Subsequently, the voltage-gated channels for the entry of Ca into the cells are unblocked and the whole process progresses. As the disease progresses, the response to flushing of more neurotransmitters is higher. We also confirmed this in our study, where we monitored the progression of changes in the visual fields in compensated glaucoma eyes over a 5-year period. We found that the greater the baseline perimeter changes were, the greater was their progression in 5 years [37].

6. The role of vascular supply of the posterior pole of the eye in hypertensive glaucoma

Another issue is that of the vascular supply to the posterior pole of the eye and its relationship to the retinal nerve fibre layer (RNFL). We also investigated the relationship between RNFL, vessel density (VD) and visual field (VF) in HTG. The results showed a moderate correlation between RNFL and VD in the same altitudinal halves of the retina. We did not find any dependence between RNFL and VF [38]. When we compared VD from the whole measured area to the sum of sensitivities in the central part of the visual field (0–22 degrees), we found a strong correlation (0.64–0.65) [39].

The relationship of VD in different stages of HTG was also studied by other authors. They all found that, with the progression of glaucoma, a reduction in VD occurs [40–45].

The IOP value also plays an important role in the value of VD. By decreasing IOP in young individuals with high IOP, Holló noted an increase in VD [46]. Conversely, after its increase above 20 mmHg, the vascular density (VD) in the macula and peripapillary decreased significantly [47].

Another important fact is the finding that glutamate can also affect blood vessels of the posterior pole of the eye. Tsuda et al. studied the effect of intravitreal NMDA in an experiment and found that NMDA induced retinal vessel loss. After blocking NMDA with nafamostat mesylate, there was a significantly lower loss of these vessels [48].

This means that, even after compensation of the IOP, the progression of changes in HTG may occur, as through the feedback, glutamate constantly bombards the neurons which have receptors for this neurotransmitter. And merely the high levels of glutamate can cause ischaemia in the eye [4–7].

Already in 1974, Hayreh [49] summarised the pathogenesis of optic disc excavation as three factors that are probably the most responsible for this abnormality:

1. Destruction of the nerve tissue in the prelaminar area;

2. Distortion of the cribriform plate (lamina cribriformis) rearwards that occurs due to retrolaminar fibrosis and a lack of normal support in the back part of the lamina due to its loss;

3. Weakening of the cribriform plate. These changes, however, are not characteristic only for optic disc glaucoma atrophy, but also have other (mainly vascular) causes.

Notwithstanding these conclusions, it is also worth noting that some local antiglaucoma medicines may lead to the progression of changes in visual fields, even following a decrease in IOP [50].

Based on this knowledge, effective antiglaucoma treatment is also offered. We emphasise the reduction of IOP in the first place. We draw attention to the unsuitability of antiglaucoma medications which can cause ischaemia of the posterior pole of the eye. This is followed by a reduction in glutamate in the synaptic cleft and blockade of its binding to NMDA receptors. The supply of energy substrates to altered nerve cells is also suitable. Because the entire visual pathway is damaged, this treatment should be systemic.

7. Conclusion

In hypertensive glaucomas, the entire visual pathway is damaged. Therefore, early diagnosis of this disease is important. Treatment should consist not only of the reduction of IOP, but also of reducing the levels of glutamate in the synaptic cleft and their binding to glutamate receptors. An important element of treatment is the supply of energy substrate to nerve cells with the ability to cope with intracellular processes. This therapy should be systemic.

Conflict of interest

The authors declare no conflict of interest.

Acronyms and abbreviations

AMPA	α-amino-3-hydroxy-5-methyl-4-8 isoxazolpropionate c/d, cup/disc ratio
CGL	corpus geniculatum laterale (lateral geniculate body)
EEAT	excitatory amino acid transporter
FMRI	functional magnetic resonance imaging
GS	glutamine synthetase
GLAST	glutamate aspartate transporter
HTG	hypertensive glaucoma;
IOP	intraocular pressure
NMDA	N-methyl-D-aspartate
MB	Muller cells
PERG	pattern electroretinogram
PET	positron emission tomography
PVEP	pattern visual evoked potential
RNFL	retinal nerve fibre layer
VD	vessel density

Author details

Jan Lestak[1,2]* and Martin Fůs[1]

1 Eye Clinic JL Faculty of Biomedical Engineering Czech Technical University in Prague, Prague, Czech Republic

2 Faculty of Biomedical Engineering, CTU in Prague, Kladno, Czech Republic

*Address all correspondence to: lestak@seznam.cz

References

[1] Lešták J, Tintěra J, Kynčl M, Svatá Z. Rozsíval P: High Tension Glaucoma and Normal Tension Glaucoma in Brain MRI J Clin Exp Ophthalmol 2013, 4: 4 DOI:10.4172/2155-9570.1000291.

[2] Tham YC, Li X, Wong TY, Quigley HA, Aung T, Cheng CY. Global prevalence of glaucoma and projections of glaucoma burden through 2040: a systematic review and meta-analysis. Ophthalmology. 2014 Nov;121(11):2081-90. DOI: 10.1016/j.ophtha.2014.05.013. Epub 2014 Jun 26. PMID: 24974815.

[3] Lešták J, Tintěra J, Kynčl M, Svatá Z. Rozsíval P: High Tension Glaucoma and Normal Tension Glaucoma in Brain MRI. J Clin Exp Ophthalmol 2013, 4:4 DOI: 10.4172/2155-9570.1000291.

[4] Sherman SM, Guillery RW. Exploring the Thalamus and Its Role in Cortical Function. 2nd Ed MIT Press; Boston: 2006.

[5] Shou TD. The functional roles of feedback projections in the visual system. Neurosci Bull. 2010 Oct;26(5):401-10. doi: 10.1007/s12264-010-0521-3. PMID: 20882067; PMCID: PMC5560353.

[6] Briggs F, Usrey WM. Corticogeniculate feedback and visual processing in the primate. J Physiol. 2011 Jan 1;589(Pt 1):33-40. doi: 10.1113/jphysiol.2010.193599. Epub 2010 Aug 19. PMID: 20724361; PMCID: PMC3039257.

[7] Thompson AD, Picard N, Min L, Fagiolini M, Chen C. Cortical Feedback Regulates Feedforward Retinogeniculate Refinement. Neuron. 2016 Sep 7;91(5):1021-1033. doi: 10.1016/j.neuron.2016.07.040. Epub 2016 Aug 18. PMID: 27545712; PMCID: PMC5156570.

[8] Kiser PD, Golczak M, Maeda A, Palczewski K. Key enzymes of the retinoid (visual) cycle in vertebrate retina. Biochim Biophys Acta. 2012 Jan;1821(1):137-51. doi: 10.1016/j.bbalip.2011.03.005. Epub 2011 Apr 5. PMID: 21447403; PMCID: PMC3158816.

[9] Clements JD, Lester RA, Tong G, Jahr CE, Westbrook GL. The time course of glutamate in the synaptic cleft. Science. 1992 Nov 27;258(5087):1498-501. doi: 10.1126/science.1359647. PMID: 1359647.

[10] Kew JN, Kemp JA. Ionotropic and metabotropic glutamate receptor structure and pharmacology. Psychopharmacology (Berl). 2005 Apr;179(1):4-29. doi: 10.1007/s00213-005-2200-z. Epub 2005 Feb 25. Erratum in: Psychopharmacology (Berl). 2005 Oct;182(2):320. PMID: 15731895.

[11] Johnson JW, Ascher P. Voltage-dependent block by intracellular Mg2+ of N-methyl-D-aspartate-activated channels. Biophys J. 1990 May;57(5):1085-90. doi: 10.1016/S0006-3495(90)82626-6. PMID: 1692749; PMCID: PMC1280813.

[12] Choi DW, Koh JY, Peters S. Pharmacology of glutamate neurotoxicity in cortical cell culture: attenuation by NMDA antagonists. J Neurosci. 1988 Jan;8(1):185-96. doi: 10.1523/JNEUROSCI.08-01-00185.1988. PMID: 2892896; PMCID: PMC6569373.

[13] Rothstein JD, Martin L, Levey AI, Dykes-Hoberg M, Jin L, Wu D, Nash N, Kuncl RW. Localization of neuronal and glial glutamate transporters. Neuron. 1994 Sep;13(3):713-25. doi: 10.1016/0896-6273(94)90038-8. PMID: 7917301.

[14] Amara SG, Fontana AC. Excitatory amino acid transporters: keeping up with glutamate. Neurochem Int. 2002 Nov;41(5):313-8. doi: 10.1016/s0197-0186(02)00018-9. PMID: 12176072.

[15] Danbolt NC. Glutamate uptake. Prog Neurobiol. 2001 Sep;65(1):1-105. doi: 10.1016/s0301-0082(00)00067-8. PMID: 11369436.

[16] Huang YH, Bergles DE. Glutamate transporters bring competition to the synapse. Curr Opin Neurobiol. 2004 Jun;14(3):346-52. doi: 10.1016/j.conb.2004.05.007. PMID: 15194115.

[17] Shen Y, Liu XL, Yang XL. N-methyl-D-aspartate receptors in the retina. Mol Neurobiol. 2006;34:163-179. doi: 10.1385/MN:34:3:163.

[18] Olney JW, Sharpe LG. Brain lesions in an infant rhesus monkey treated with monsodium glutamate. Science. 1969 Oct 17;166(3903):386-8. doi: 10.1126/science.166.3903.386. PMID: 5812037.

[19] Vorwerk CK, Naskar R, Dreyer EB. Die exzitotoxische Hypothese der Glaukomgenese [The excitotoxicity theory of glaucoma]. Klin Monbl Augenheilkd. 1999 Jan;214(1):2-11. German. doi: 10.1055/s-2008-1034740. PMID: 10198875.

[20] Woldemussie E, Wijono M, Ruiz G. Müller cell response to laser-induced increase in intraocular pressure in rats. Glia. 2004 Aug 1;47(2):109-19. doi: 10.1002/glia.20000. PMID: 15185390.

[21] Grewer C, Gameiro A, Zhang Z, Tao Z, Braams S, Rauen T. Glutamate forward and reverse transport: from molecular mechanism to transporter-mediated release after ischemia. IUBMB Life. 2008 Sep;60(9):609-19. doi: 10.1002/iub.98. PMID: 18543277; PMCID: PMC2632779.

[22] Pavlidis M, Stupp T, Naskar R, Cengiz C, Thanos S. Retinal ganglion cells resistant to advanced glaucoma: a postmortem study of human retinas with the carbocyanine dye DiI. Invest Ophthalmol Vis Sci. 2003 Dec;44(12):5196-205. doi: 10.1167/iovs.03-0614. PMID: 14638717.

[23] Shou T, Liu J, Wang W, Zhou Y, Zhao K. Differential dendritic shrinkage of alpha and beta retinal ganglion cells in cats with chronic glaucoma. Invest Ophthalmol Vis Sci. 2003 Jul;44(7):3005-10. doi: 10.1167/iovs.02-0620. PMID: 12824245.

[24] Morgan JE, Uchida H, Caprioli J. Retinal ganglion cell death in experimental glaucoma. Br J Ophthalmol. 2000 Mar;84(3):303-10. doi: 10.1136/bjo.84.3.303. PMID: 10684843; PMCID: PMC1723413.

[25] Naskar R, Wissing M, Thanos S. Detection of early neuron degeneration and accompanying microglial responses in the retina of a rat model of glaucoma. Invest Ophthalmol Vis Sci. 2002 Sep;43(9):2962-8. PMID: 12202516.

[26] Soto I, Oglesby E, Buckingham BP, Son JL, Roberson ED, Steele MR, Inman DM, Vetter ML, Horner PJ, Marsh-Armstrong N. Retinal ganglion cells downregulate gene expression and lose their axons within the optic nerve head in a mouse glaucoma model. J Neurosci. 2008 Jan 9;28(2):548-61. doi: 10.1523/JNEUROSCI.3714-07.2008. PMID: 18184797; PMCID: PMC6670511.

[27] Lestak J, Tintera J, Svata Z, Ettler L, Rozsival P. Glaucoma and CNS. Comparison of fMRI results in high tension and normal tension glaucoma. Biomed Pap Med Fac Univ Palacky Olomouc Czech Repub. 2014;158(1):144-53. doi: 10.5507/bp.2013.038. Epub 2013 Jun 4. PMID: 23733146.

[28] Fortune B, Bui BV, Morrison JC, Johnson EC, Dong J, Cepurna WO, Jia L, Barber S, Cioffi GA. Selective ganglion cell functional loss in rats with experimental glaucoma. Invest Ophthalmol Vis Sci. 2004 Jun;45(6):1854-62. doi: 10.1167/iovs.03-1411. PMID: 15161850.

[29] Holder GE. Pattern electroretinography (PERG) and an

integrated approach to visual pathway diagnosis. Prog Retin Eye Res. 2001 Jul;20(4):531-61. doi: 10.1016/s1350-9462(00)00030-6. PMID: 11390258.

[30] Parisi V, Miglior S, Manni G, Centofanti M, Bucci MG. Clinical ability of pattern electroretinograms and visual evoked potentials in detecting visual dysfunction in ocular hypertension and glaucoma. Ophthalmology. 2006 Feb;113(2):216-28. doi: 10.1016/j.ophtha.2005.10.044. Epub 2006 Jan 10. PMID: 16406535.

[31] NEBBIOSO, Marcella, Fabio De GREGORIO, Laura PRENCIPE a Irene PECORELLA. Psychophysical and Electrophysiological Testing in Ocular Hypertension. Optometry and Vision Science [online]. 2011, 88(8), E928-E939 [cit. 2020-12-30]. ISSN 1040-5488. Dostupné z: doi:10.1097/OPX.0b013e31821c6ca4

[32] Lešták, J., Nutterová, E., Pitrová, Š., Krejčová, H., Bartošová, L., Forgáčová, V.: High tension versus normal tension glaucoma. A comparison of structural and functional examinations. J Clinic Exp Ophthalmol 2012, S:5, doi: 10.4172/2155-9570.S5-006.

[33] Castro NG, de Mello MC, de Mello FG, Aracava Y. Direct inhibition of the N-methyl-D-aspartate receptor channel by dopamine and (+)-SKF38393. Br J Pharmacol. 1999 Apr;126(8):1847-55. doi: 10.1038/sj.bjp.0702479. PMID: 10372829; PMCID: PMC1565957.

[34] Wu Y, Pearl SM, Zigmond MJ, Michael AC. Inhibitory glutamatergic regulation of evoked dopamine release in striatum. Neuroscience. 2000;96(1):65-72. doi: 10.1016/s0306-4522(99)00539-4. PMID: 10683411.

[35] Kaneko M, Sugawara T, Tazawa Y. [Electrical responses from the inner retina of rats with streptozotocin-induced early diabetes mellitus]. Nippon Ganka Gakkai Zasshi. 2000 Nov;104(11):775-8. Japanese. PMID: 11530366.

[36] Lestak J, Fus M: Neuroprotection in glaucoma – a review of electrophysiologist. Exp Ther Med. 2020 Apr;19(4):2401-2405. doi: 10.3892/etm.2020.8509. Epub 2020 Feb 10

[37] Lešták, J., Rozsíval, P.: The Influence of Corneal Thickness on Progression of Hypertensive Glaucoma. J Clin Exp Ophthalmol 2012, 3:8, doi: 10.4172/2155-9570.1000245.

[38] Zakova M, Lestak J, Fus M, Maresova K. OCT angiography and visual field in hypertensive and normotensive glaucomas. Biomed Pap Med Fac Univ Palacky Olomouc Czech Repub.2020, 164, DOI: 10.5507/bp.2020.044

[39] Lešták J, Fůs M, Benda A, Marešová K. OCT ANGIOGRAPHY AND DOPPLER SONOGRAPHY IN NORMAL-TENSION GLAUCOMA. Ceska a Slovenska Oftalmologie: Casopis Ceske Oftalmologicke Spolecnosti a Slovenske Oftalmologicke Spolecnosti. 2020 ;76(3):120-123. DOI: 10.31348/2020/20.

[40] Mammo Z, Heisler M, Balaratnasingam C, et al. Quantitative Optical Coherence Tomography Angiography of Radial Peripapillary Capillaries in Glaucoma, Glaucoma Suspect, and Normal Eyes. Am J Ophthalmol. 2016;170:41-49.

[41] Yarmohammadi A, Zangwill LM, Diniz-Filho A, Suh MH, Yousefi S, Saunders LJ, Belghith A, Manalastas PI, Medeiros FA, Weinreb RN. Relationship between Optical Coherence Tomography Angiography Vessel Density and Severity of Visual Field Loss in Glaucoma. Ophthalmology. 2016 Dec;123(12):2498-2508.

[42] Hou H, Moghimi S, Zangwill LM, Shoji T, Ghahari E, Manalastas PIC,

Penteado RC, Weinreb RN. Inter-eye Asymmetry of Optical Coherence Tomography Angiography Vessel Density in Bilateral Glaucoma, Glaucoma Suspect, and Healthy Eyes. Am J Ophthalmol. 2018;190:69-77.

[43] Hollo G. Comparison of Peripapillary OCT Angiography Vessel Density and Retinal Nerve Fiber Layer Thickness Measurements for Their Ability to Detect Progression in Glaucoma. J Glaucoma. 2018;27:302-305.

[44] Penteado RC, Zangwill LM, Daga FB, Saunders LJ, Manalastas PIC, Shoji T, AkagiT,ChristopherM,YarmohammadiA, Moghimi S, Weinreb RN. Optical Coherence Tomography Angiography Macular Vascular Density Measurements and the Central 10-2 Visual Field in Glaucoma. J Glaucoma. 2018;27:481-489.

[45] Mangouritsas G, Koutropoulou N, Ragkousis A, Boutouri E, Diagourtas A. Peripapillary Vessel Density In Unilateral Preperimetric Glaucoma. Clin Ophthalmol. 2019;13:2511-2519. doi: 10.2147/OPTH.S224757.

[46] Holló G. Influence of Large Intraocular Pressure Reduction on Peripapillary OCT Vessel Density in Ocular Hypertensive and Glaucoma Eyes. J Glaucoma. 2017 Jan;26(1):e7-e10. doi: 10.1097/IJG.0000000000000527.

[47] Ma ZW, Qiu WH, Zhou DN, Yang WH, Pan XF, Chen H. Changes in vessel density of the patients with narrow antenior chamber after an acute intraocular pressure elevation observed by OCT angiography. BMC Ophthalmol. 2019 Jun 21;19(1):132. doi: 10.1186/s12886-019-1146-6

[48] Uchiba M, Okajima K, Murakami K, Okabe H, Takatsuki K. Effect of nafamostat mesilate on pulmonary vascular injury induced by lipopolysaccharide in rats. Am J Respir Crit Care Med. 1997 Feb;155(2):711-8. doi: 10.1164/ajrccm.155.2.9032217. PMID: 9032217.

[49] Hayreh SS. Pathogenesis of cupping of the optic disc. Br J Ophthalmol. 1974 Oct;58(10):863-76. doi: 10.1136/bjo.58.10.863. PMID: 4375487; PMCID: PMC1215042.

[50] Maresova K, Lestak J, Fus M, Weisova, Effect of prostaglandins and beta blockers on progression of hypertensive and normotensive glaucoma. Biomed Pap Med Fac Univ Palacky Olomouc Czech Repub 2020 Mar 26. doi: 10.5507/bp.2020.011

Section 2

Miscellaneous

Chapter 5

Psychosocial Aspects of Glaucoma

Ashutosh Dayal

Abstract

Glaucoma, not only leads to irreversible visual impairment, but also has a negative impact on Quality of Life of the patients. Visual disability, lifelong medical and surgical treatments and even the mere knowledge of having an irreversible potentially blinding condition tend to cause severe psychological stress in patients, leading to negative emotions such as anxiety or depression. The goal of glaucoma treatment should not be limited to preserving the vision, but also address the psychological aspects and preservation of patient wellbeing. Patient counselling, right from at the time of diagnosis, periodic psychological assessment and creating awareness in the society as a whole should be implemented as a part of holistic approach to glaucoma. Utilisation of Patient- reported quality of life tools would help clinicians in more closely understanding the problems and would, in turn, aid in providing comprehensive customised treatment option for each patient.

Keywords: Glaucoma, psychosocial impairment, Quality of life, Patient counselling

1. Introduction

Glaucoma is a chronic progressive optic nerve disease that can potentially lead to blindness, if left untreated. It is the second most common cause of blindness and the leading cause of irreversible blindness worldwide [1]. Global prevalence of glaucoma has been estimated to be 3.54%, though geographic variations exist. Incidence of Primary open angle glaucoma has been found to be highest in African countries, while that of primary angle closure glaucoma in Asian populations [2]. Presently, 80 million people are affected with glaucoma, worldwide. This number is expected to rise to almost 111 million by the year 2040 [2, 3]. However, these numbers may just be reflecting the tip of the iceberg. Owing to slow and silent progression, the disease remains undiagnosed in many individuals till fairly advanced stages. Diagnosis is often delayed and approximately 40% of retinal nerve fibres are lost by the time field defects could be appreciated on standard White-on-White perimetry [4]. Moreover, the disease characteristically involves peripheral and mid peripheral visual fields initially, which may go unnoticed in many patients. Elderly population is at a greater risk as they tend to ignore it as an age-related inevitability [5]. Unfortunately, at the time of first presentation, most patients have significant irreversible peripheral visual field defects. This is truer for population residing in developing and underdeveloped countries, with less awareness and limited access to basic eyecare services. Some of the earlier epidemiologic studies conducted in West Indies, [6] Australia [7] and Netherlands [8] reported more than 50% patients having undiagnosed glaucoma during screening. Data from Indian studies reported an even higher proportion of undiagnosed cases discovered during screening [9–11].

World health organisation (WHO) has defined Quality of life as "individuals perception of their position in life in the context of the culture and value systems in which they live and in relation to their goals, expectations, standards and concerns." [12]. In other words, it can be defined as an individual's own measure of various life conditions experienced by him, including physical health, mental and psychological health, dependency, social functioning and economic wellbeing. Thus, conditions affecting any or all of the above-mentioned aspects, would hamper Quality of life. Glaucoma not only causes permanent visual disability, but also has a negative influence on mental and psychological health, which is subjective and individual specific [5]. Individuals with glaucoma tend to develop negative emotions such as anxiety or depression, which are detrimental in the daily functioning and wellbeing. With the number of glaucoma cases projected to rise, the psychosocial aspect of glaucoma is now becoming a global concern for clinicians all over.

2. Factors affecting quality of life in glaucoma

2.1 Disability due to visual field defects

Glaucoma causes irreversible defects in the visual fields, that are slow and progressive. Glaucomatous visual field defects appear to be non-specific and follow the pattern of retinal nerve fibre layer loss [13]. Initially, the disease involves peripheral fields and tends to go unnoticed, till the time it progresses to involve central and paracentral areas. Hence, it is less common for an individual with early glaucoma to present with visual symptoms. Although, some individuals, particularly those involved in driving or sports, may notice these peripheral field defects rather early. Typically, by the time a patient presents with visual symptoms, significant field defects have already set in. With progression of the disease, patients find it more and more difficult to pursue day to day activities. They find themselves bumping into objects while walking, as they are unable to perceive obstacles in the path which lie in the non-seeing areas of the visual field. In advance stages, blindness ensues causing complete dependency even for performing routine tasks. Studies have demonstrated that beside visual fields, glaucoma also affects other visual functions such as contrast sensitivity and dark adaptation. Impairment of these functions can further compound the problems for the patients. Activities such as night time driving, that require good peripheral vision with dark adaptation and contrast sensitivity, become difficult to execute. All these factors cause significant visual disability among patients, even in moderate stages of the disease. Difficulty in carrying out daily activities and fear of falling increases dependency and limits social functioning [14]. This causes a negative effect on patients' thought process, and directly impacts their mental health. It is evident through various studies, that deterioration of Quality of life and mental health status is associated with the worsening of visual impairment in glaucoma [15–17].

2.2 Adverse effect related to treatment

Intraocular pressure is the only known modifiable risk factor known for glaucoma [18]. Hence the current treatment options aim at reducing the intraocular pressure. This can be achieved either by medications, laser treatment or filtration surgeries. Conventionally medical management, by the use of topical or systemic intraocular pressure lowering agents, is considered as the first line therapy. Monotherapy with one agent or a combination of two or more agents may be required to achieve the target intraocular pressure, depending on the severity of disease. Here we briefly discuss some of the commonly prescribed topical medication and related adverse effects:

Prostaglandin Analogues (e.g., Latanoprost, Bimatoprost, Travoprost): Preferred as first line agents in glaucoma patients. They lower the intraocular pressure most effectively, with once daily dosing, by facilitating the uveoscleral outflow. Prostaglandin analogues do have several associated ocular side effects including darkening of iris colour, trichomegaly (eyelash growth), periocular skin pigmentation, conjunctival hyperaemia and cystoid macular edema.

Beta adrenergic blockers (e.g., Timolol, Betaxolol, Levobunolol): One of the most commonly used drugs as first line therapy. It reduces the production of aqueous humour by acting on the ciliary body, inhibiting beta adrenergic receptors. Effects may be lost in time due to tachyphylaxis. In addition, it can be associated with local and systemic side effects such as bradycardia, bronchospasm, hypoglycaemia, depression, dry eye, punctate keratitis. Ocular stinginess is associated with betaxolol.

Alpha adrenergic receptor agonist (e.g.: Brimonidine): It is potent class of antiglaucoma agents used as first line therapy. It is recommended mainly for chronic use in patients with concomitant cardiopulmonary disease or any other contraindications to beta-blockers. Adverse effects include allergic conjunctivitis, conjunctival hyperaemia, pruritis, dry eye, fatigue and drowsiness.

Carbonic Anhydrase inhibitor (e.g.: Dorzolamide, Brinzolamide): It is derivative of sulphonamide, which decreases the production of aqueous humour. After corneal penetration, it inhibits the carbonic anhydrase in the ciliary body, slowing local bicarbonate production which in turn decreases sodium and fluid transport, thus reducing aqueous humour secretion. It can precipitate corneal edema with patients with decompensated cornea. Other ocular adverse effects include stinging sensation, metallic taste in mouth after instillation and granulomatous anterior uveitis (rare).

All the topical agents have potential to cause localised ocular adverse effects that can cause discomfort to the patient. Moreover, many of the intraocular pressure lowering agents require more than once daily dosing schedule, that elderly patients might find tedious or difficult to follow. Apart from the ocular and systemic effects of the topical medications, problems associated with preservatives present in the preparations, on ocular surface, should be borne in mind. Preservatives are additive agents added to extend the shelf-life of an ophthalmic preparation. Commonly used preservatives include Benzalkonium chloride (BAK), Purite, parabens, chlorobutanol, sodium chlorite, and a boric acid plus D-sorbitol plus zinc chloride fixed combination. However, these preservative agents are known to destabilise the cell membrane, causing cell loss and reduced cell surface adhesions, in normal corneal and conjunctival tissue. This may give rise to various ocular surface diseases, including superficial punctate keratitis with corneal erosions, papillary conjunctivitis, dry eye, low tear film break-up time, conjunctival injection and anterior chamber inflammation.

Treatment for glaucoma requires lifelong instillation of intraocular pressure lowering medications. Chronic instillation of these medications over a period of time can accentuate insult to the ocular surface. Recent studies have demonstrated cellular level changes, including epithelial thickening, loss of goblet cells and conjunctival stromal thickening, associated with long term use of antiglaucoma medications [19]. With altered ocular surface structure, the tear film function gets compromised resulting in conditions such as dry eye syndrome and conjunctivitis medicamentosa. Prevalence of ocular surface disease has been estimated to be 59% in patients with glaucoma, with medication playing a significant part [20]. Ocular surface disease was found to have a negative impact on patient reported quality of life scores, with majority reporting ocular pain and discomfort [16, 21]. Although preservative free preparations are now available in the market, which claim to be free from adverse effects of preservatives, further studies would be required to conclusively prove their superiority over the conventional preparations.

On the other hand, surgical interventions such as trabeculectomy or drainage device implant, though reduce the dependency on topical medications, come with their own disadvantages. These include intraocular pressure fluctuations, lifelong follow ups, need for secondary interventions such as antimetabolite injections, bleb needling or repeat surgery in case of filtration failure. Patients with surgical intervention reported more ocular discomfort, probably owing to the presence of conjunctival bleb on the ocular surface [22].

2.3 Psychological impact of diagnosis of glaucoma

As with other chronic diseases, diagnosis of glaucoma can be a stressful experience and a psychological burden for the patient. Even on just learning about the diagnosis, many patients develop negative emotions such as anxiety or depression. Given the potential blinding nature of the disease, bulk of the negative thoughts arise due to the fear of going blind. Moreover, factors such as changes in daily routine, regular use of medications, lifelong treatment and follow ups can cause psychological stress. Studies done in recent decades have provided us with an insight into the patient thought process. In the Collaborative Initial Glaucoma treatment study, the newly diagnosed patients were found to have moderate to severe psychological fear of blindness at the time of diagnosis [23]. Odberg et al. reported negative emotions in almost 90% of newly diagnosed cases of glaucoma [24]. This psychological stress may be partly attributed to lack of awareness about the disease in general population. Patients tend to misinterpret the diagnosis of glaucoma as that of impending blindness. They are totally unaware of the treatment options available, and the fact that the disease progression can be retarded or halted with the use of various medical and surgical modalities. Poor disease comprehension has shown to be negatively associated with psychological and quality of life scores [25]. It should be borne in mind that patients can develop psychological emotions such as anxiety and fear of potential outcomes irrespective of duration of disease.

2.4 Financial and economic perspectives

Glaucoma, being a chronic disease, can become a financial burden for the patients. Cumulative cost of treatment over many years- whether medical or surgical- can be a major problem for patients. A significant proportion of patients reside in middle to low-income countries, with low per capita income and lack of government funded social security schemes, affordability of treatment may be a challenge for them in the long run. These patients have to spend a large share of their monthly income on treatment for glaucoma. A single follow up visit to the hospital leads to financial loss to those employed as daily wage workers. For people residing in remote areas with inadequate health facilities, even travelling all the way for a follow-up visit is a herculean task. All these factors play a significant role in patients not adhering to treatment and follow-ups. As the disease progresses and visual fields deteriorate, patients find themselves at a constant risk of loss of livelihood. With loss of livelihood, and limited or no other sources of income, continuing treatment of glaucoma becomes virtually impossible.

Hence it is evident that glaucoma associated psychosocial impairment is multifactorial, and not just caused by loss of visual function alone. Psychological and financial factors play an equal role in determining the quality of life of a patient. In recent decades, much attention has been drawn towards assessment of the psychological effects of glaucoma, in an attempt to preserve social wellbeing of the patients. Population based cross sectional studies have been conducted in both developed as well as developing and underdeveloped regions. Studies conducted in Singapore, [26] China, [25] Japan, [17] India, [15, 16] Germany [27] and Nigeria [28] have brought forth the conclusion that glaucoma does have a deleterious effect on patient psychology, mental health status, social functioning and above all quality of life. The quality of life of deteriorates as the disease progresses, owing to worsening of visual fields, limitation of social functioning and increased dependency. Significant prevalence of anxiety and depression in patients with glaucoma reflects the enormous psychological burden of the disease [26, 27].

3. Management- what needs to be done?

Earlier, management for glaucoma was aimed only at preserving the visual function by control of intraocular pressure. Less emphasis was given to psychological impact of the disease as well as its implication on quality of life of the patients. Today, with the data from various studies, we now know the mental, social, financial and psychological implications of the disease. Considering the prevalence of glaucoma, the magnitude of problem is tremendous. Hence, mere treatment of raised intraocular pressure may not be sufficient. One should aim at providing a decent quality of life and preserving patient wellbeing. It should be borne in mind that one should treat the patient and not just intraocular pressure.

3.1 Patient counselling

It has been demonstrated that disease comprehension is often poor among patients, causing psychological stress and feeling of despair and panic. Counselling of patients about the disease can help alleviate this problem to some extent. Counselling should be started right at the time of diagnosis, by the treating clinician and continued on periodic basis over the course of treatment. Patients should be made aware about the disease, treatment options available for them, importance of treatment compliance and periodic follow-up. This would help in alleviation the psychological fear, as shown by Odberg et al. in their study [24]. During the course of treatment, patients may require periodic motivation to adhere to the treatment and follow ups.

3.2 Assessment of psychological state and quality of life

A qualitative or quantitative assessment of psychological state of the patient should be included as a part of glaucoma management. This can be performed with the use of various tools in the form questionnaires. Some of the commonly available tools are mentioned in **Table 1**. These questionnaires can be either generic (Short Form-36) or vision specific (NEIVFQ-51, NEIVFQ-25) or disease specific (Glaucoma Quality of Life –15, Glaucoma symptom scale). Although any of the above-mentioned tools can be utilised as per convenience, Patient reported outcomes (PRO) have been more commonly used tools. PRO based questionnaires provide a better perspective of the difficulties experienced by the patient. Apart from quality of life, separate instruments are available for measuring negative mental conditions-precisely anxiety and depression. Some of the tools such as Hospital Anxiety and Depression scale (HADS), General Anxiety Disorder-7 (GAD-7) are easy to use and can be self-administered in the clinic or at home. As with counselling, such assessment should be done at the time of diagnosis and periodically during follow ups. Any deterioration in the scores would warrant need for intervention such as counselling.

Name of questionnaire	Brief Description
National eye institute visual function questionnaire −25 (NEIVFQ-25)	25 -item questionnaire measuring vision dependent functioning and influence of vision problems on quality of life. Measures various patient reported functions including general health, mental health, social health, dependency, driving, near vision, colour vision, ocular pain etc. All these sub scores are used to calculate composite score. It measures Patient reported outcomes (PRO)
Glaucoma Quality of Life −15 (GQL-15)	It's a disease specific questionnaire. Assesses only vision related difficulty, taking into account the effect of binocular visual field loss on visual function.
Short Form-36 (SF 36)	Short general health questionnaire. Assesses the physical component, mental health, emotional- role and social functioning. It was used to measure quality of life in Ocular hypertension treatment trial.
World Health organisation BREF (WHO-BREF)	It is the shorter version of the original 100 item questionnaire. Measures General health, positive feeling, social support, financial resources/ physical, psychological and social relationships

Table 1.
Brief description of some commonly used validates questionnaires for assessment of quality of life.

It should be kept in mind that individuals with poor quality of life scores or negative emotions are susceptible to treatment non-compliance. Hence it is essential that the treating clinicians remain aware of the psychological condition of their patients.

3.3 Vocational training

Visual disability becomes profound as the disease progresses to advanced stages. Apart from causing limitations in social functioning, it can lead to loss of livelihood. This may be detrimental to patient's quality of life causing low self-esteem and frustration. In such scenarios, vocational training and use of low vision aid devices can be beneficial for many patients. Use of low vision devices such as magnifiers, telescopes or field expanders can help patients in utilising their residual field of vision. Vocational training activities would help them with performing routine tasks independently. Moreover, they can learn newer skills that may generate livelihood opportunities for some. These activities can provide patients with much needed self-confidence, reduce dependency and would aid in improving Quality of life to some extent.

3.4 Creating awareness among general population

With constant rise in number of people with glaucoma, there is a need to create awareness about the disease among the masses. Sensitising the community regarding the disease, its treatment as well as about the problems faced by the patients would help in generating community- based support for patients with glaucoma. Such moral support from friends and family members would be beneficial for the patients in keeping themselves motivated while facing the challenges associated with glaucoma. Community based social groups for persons with glaucoma should be promoted, wherein patients can share their experiences and problems.

4. Conclusion

Patients with glaucoma are prone for developing psychological disturbances that in tun negatively affect quality of life. Apart from visual disability, psychological effect of diagnosis, treatment related adverse effects and financial issues play an important role in hampering the quality of life. A comprehensive approach consisting of patient education, psychological assessment, motivational counselling and vocational therapy should be adopted as a part of glaucoma management protocol. This would enable clinicians to provide customised holistic treatment for each patient, thereby increasing compliance and providing better quality of life.

Conflict of interest

The authors declare no conflict of interest.

Author details

Ashutosh Dayal
Department of Glaucoma Services, Sankara Nethralaya (Medical Research Foundation), Chennai, India

*Address all correspondence to: adayal007@gmail.com

References

[1] World Health Organization.Global data on visual impairment 2010 [Internet].2012. Available from: https://www.who.int/blindness/GLOBAL DATAFINALforweb.pdf [Last accessed 2021-03-07]

[2] Tham Y-C, Li X, Wong TY, Quigley HA, Aung T, Cheng C-Y. Global Prevalence of Glaucoma and Projections of Glaucoma Burden through 2040. Ophthalmology. 2014;121(11):2081-90.

[3] Pascolini D, Mariotti SP. Global estimates of visual impairment: 2010. Br J Ophthalmol.2012;96(5):614-8.

[4] Michelessi M, Lucenteforte E, Oddone F, Brazzelli M, Parravano M, Franchi S, et al. Optic nerve head and fibre layer imaging for diagnosing glaucoma. Cochrane database Syst Rev.2015;11(11):CD008803.

[5] Kaur D, Gupta A, Singh G. Perspectives on Quality of Life in Glaucoma. J Curr glaucoma Pract 2012;6(1):9-12.

[6] Leske MC, Connell AMS, Schachat AP. The Barbados Eye Study: Prevalence of Open Angle Glaucoma. Arch Ophthalmol.1994;112(6):821-9.

[7] Ivers RQ, Cumming RG, Mitchell P, Attebo K. Visual impairment and falls in older adults: the Blue Mountains Eye Study. J Am Geriatr Soc. 1998;46(1):58-64.

[8] Dielemans I, Vingerling JR, Wolfs RCW, Hofman A, Grobbee DE, de Jong PTVM. The Prevalence of Primary Open-angle Glaucoma in a Population-based Study in The Netherlands: The Rotterdam Study. Ophthalmol ogy.1994;101(11):1851-5.

[9] Vijaya L, George R, Baskaran M, Arvind H, Raju P, Ramesh SV, et al. Prevalence of Primary Open-angle Glaucoma in an Urban South Indian Population and Comparison with a Rural Population. Ophthalmology. 2008;115(4):648-654.

[10] Ramakrishnan R, Nirmalan PK, Krishnadas R, Thulasiraj RD, Tielsch JM, Katz J, et al. Glaucoma in a rural population of Southern India: The Aravind Comprehensive Eye Survey. Ophthalmology. 2003;110(8):1484-90.

[11] Dandona L, Dandona R, Srinivas M, Mandal P, John RK, McCarty CA, et al. Open-angle glaucoma in an urban population in southern India: the Andhra Pradesh eye disease study. Ophthalmology [Internet]. 2000;107(9):1702-9.

[12] Division of Mental Health and Prevention of Substance Abuse WHO. Measuring quality of life. 1997. Available from http://www.who.int/mental_health/media.68.pdf [Last accessed 2021-03-10]

[13] Broadway DC. Visual field testing for glaucoma - a practical guide. Community eye Heal.2012;25:66-70.

[14] Freeman EE, Mun~oz B, Rubin G, West SK. Visual Field Loss Increases the Risk of Falls in Older Adults: The Salisbury Eye Evaluation. Investig Opthalmology Vis Sci. 2007;48(10):4445.

[15] Kumar S, Ichhpujani P, Singh R, Thakur S, Sharma M, Nagpal N. The impact of primary open-angle glaucoma: Quality of life in Indian patients. Indian J Ophthalmol. 2018;66(3):416-9.

[16] Kalyani VKS, Dayal A, Chelerkar V, Deshpande M, Chakma A. Assessment of psychosocial impact of primary glaucoma and its effect on quality of life of patients in Western India. Indian J Ophthalmol.2020;68(11):2435-8.

[17] Sawada H, Fukuchi T, Abe H. Evaluation of the relationship between quality of vision and the visual function index in Japanese glaucoma patients. Graefe's Arch Clin Exp Ophthal mol.2011;249(11):1721-7.

[18] Kass MA, Heuer DK, Higginbotham EJ, Johnson CA, Keltner JL, Miller JP, et al. The Ocular Hypertension Treatment Study: a randomized trial determines that topical ocular hypotensive medication delays or prevents the onset of primary open-angle glaucoma. Arch Ophthalmol (Chicago, Ill 1960).2002;120(6):701-13

[19] Mastropasqua L, Agnifili L, Mastropasqua R, Fasanella V. Conjunctival modifications induced by medical and surgical therapies in patients with glaucoma. Curr Opin Pharmacol.2013;13(1): 56-64.

[20] Leung EW, Medeiros FA, Weinreb RN. Prevalence of ocular surface disease in glaucoma patients. J Glaucoma.2008;17(5):350-5.

[21] Rossi GCM, Pasinetti GM, Scudeller L, Bianchi PE. Ocular surface disease and glaucoma: How to evaluate impact on quality of life. J Ocul Pharmacol Ther. 2013;29(4):390-4

[22] Janz NK, Wren PA, Lichter PR, Musch DC, Gillespie BW, Guire KE, et al. The collaborative initial glaucoma treatment study: Interim quality of life findings after initial medical or surgical treatment of glaucoma. Ophthalmology. 2001;108(11):1954-65.

[23] Jampel HD, Frick KD, Janz NK, Wren PA, Musch DC, Rimal R, et al. Depression and Mood Indicators in Newly Diagnosed Glaucoma Patients. Am J Ophthalmol. 2007;144(2):238-44

[24] Odberg T, Jakobsen JE, Hultgren SJ, Halseide R. The impact of glaucoma on the quality of life of patients in Norway. I. Results from a self-administered questionnaire. Acta Ophthalmol Scand.2001 ;79(2):116-20.

[25] Mei Kong X, Qing Zhu W, Xu Hong J, Huai Sun X. Is glaucoma comprehension associated with psychological disturbance and vision-related quality of life for patients with glaucoma? A cross-sectional study. BMJ Open. 2014;4:4632.

[26] Lim NCS, Fan CHJ, Yong MKH, Wong EPY, Yip LWY. Assessment of Depression, Anxiety, and Quality of Life in Singaporean Patients with Glaucoma. J Glaucoma. 2016;25(7):605-12.

[27] Rezapour J, Nickels S, Schuster AK, Michal M, Münzel T, Wild PS, et al. Prevalence of depression and anxiety among participants with glaucoma in a population-based cohort study: The Gutenberg Health Study. BMC Ophthalmol. 2018;18(1):157

[28] Onwubiko SN, Nwachukwu NZ, Muomah RC, Okoloagu NM, Ngwegu OM, Nwachukwu DC. Factors Associated with Depression and Anxiety among Glaucoma Patients in a Tertiary Hospital South-East Nigeria. Niger J Clin Prac. 2020;23:315-21

Chapter 6

Physiological Bases of Electric Stimulation as a New Approach to Glaucoma IOP Control

Luis Nino-de-Rivera, Diego Cervera and Paola Castillo-Juarez

Abstract

This Chapter focuses in the electrophysiological bases to support Trans Palpebral Electrical Stimulation TPES as a new alternative to control Intraocular Pressure IOP. Primary open Angle Glaucoma POAG is described in our approach as a dysfunction of the membrane potential of TM cells due to the dysfunction of the Maxi potassium depended Calcium Channels $BKCa_2+$ of the Trabecular Mesh TM. We review through the paper the main contributions about Trabecular mesh dysfunction related with Voltage dependent ionic channels. We also present in this paper new results in controlling intra ocular pressure IOP during one year of trans palpebral Electric stimulation in patients with Primary open-angle glaucoma (POAG).

Keywords: Glaucoma, Trans Palpebral Electrical Stimulation TPES, IOP

1. Introduction

Glaucoma is the first cause of blindness all over the world [1]. The main risk of this disease is an increase in intraocular pressure (IOP); this is due to the dysfunction of the trabecular mesh that does not let ocular drainage flux properly [2]. Most POAG treatments go to lower IOP decreasing the production of the aqueous humor (AH), however they do not go to the origin of the problem: the dysfunction of the trabecular mesh to control aqueous (AH) flux system [3–6]. We show in **Figure 1** an approach of the IOP regulation mechanism, regulated by the flux that goes among TM cells. Cells have the capacity to grow or decrease their volumes adaptively, regulating the AH to go out. Consider the dysfunction of the conventional drainage of the TM cells if they lose the capacity to regulate their volume avoiding the right passage of intraocular flux among them. The Trabecular mash cells work like inside a tube, they let the aqueous humor pass amid the cells, then the AH goes from top to bottom passing trough the free spaces the cells liberate. The control of the AH flux is like balls in a tube varying their volume. The function of the TM cells is to regulate the passage of the humor through the free spaces depending on its volume. The highest cell volume, the lower flux to circulate among the balls, then higher pressure over inside the ocular globe surface. Ionic channels are device like swivels that allow to liberate pressure. Once the ionic channels fails closing abnormally, TM Cells do not control the AH passage among the *"cell-ball"*. Inside the ball there is a continuous production of water and K+ ions, if no liberation way of them, the cell grows in volume, increasing the pressure over the eye globe wall.

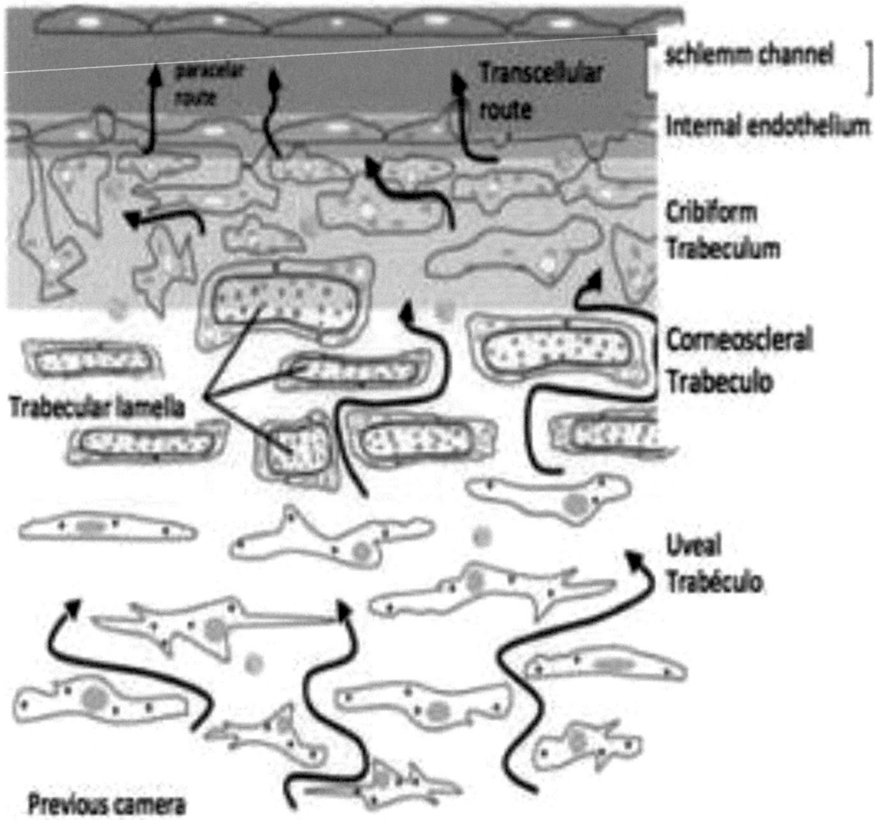

Figure 1.
The scheme shows how aqueous humor AH, shown in the scheme like arrows, pass thought out TM cells. TM cells regulates the AH flux varying their volume in order to control the amount of AH flux.

The unfixed regulation on the opened-closed processes of depended voltage ionic channels ($BKCa_2+$) affects the volume regulation of the cell. This inflammatory process regulation is related with Voltage-dependent $BKCa_2+$ channels VDCH which deregulation plays a central role in POAG illness [7–10]. BKCa2 ionic channels works like a hinge closing (holy or partially) to regulate the positive charge inside cell due to an increase of Ca + inside the TM cells. BKCa2+ works as a control system to regulate the positive charge ions inside the cell and then regulating the elasticity, contractility and then the volume of the cell. $BKCa_2+$ regulates the positive charge inside the cell regulating the cell membrane potential.

2. Glaucoma and TM ion channel dysfunction

Stumpff and Soto [8, 9] reported that some tyrosine kinase in the trabecular mesh can capture the regulation function of the $BKCa_2+$ channels, avoiding its volume regulation. Some Kinases avoids VDCH to act as a hinge regulating the volume of the cell capturing the Voltage sensor of the BKCa2+ sensor. The deregulation of the volume of the cell is, due to a dysfunction of some ionic channels is a typical channelopathy, it goes to an uncontrolled typical inflammatory process. TM cell dysfunction goes cell to its highest volume, losing the ability on going back them to a lower volume. This lost of flexibility shows: a) The incapacity of the BKC_a+ channel to open and consequently to liberate H_2O plus K+, as required.

B) The deregulation of the membrane potential $V_{m,}$, is due to an excess of positive charge (+ch) inside the cytosol, not properly liberated by the BKC_a+ channels, this is equivalent to keep closed the VDCH channel by keeping the membrane potential in the depolarization region avoiding the liberation of (+ch) and water. The right regulation of charges in the cell and its signaling (+ch) depends on the right V_m swing performance. A key point on this is how the V_m properly swings. We illustrate in **Figure 2a** the membrane potential $V_{m,.}$ The depolarization region shows the +charge (+ch) growing inside the cell, this region goes from the equilibrium or rest voltage V_r to its highest positive value. The repolarization region occurs when the channel opens liberating K+. It goes from the highest V_m positive voltage (the highest positive charge allowed inside the cell), to the rest voltage V_r. This new approach to understand glaucoma let us to think in new solutions to force the cell membrane to get the right potential by an external voltage.

Moreover, Stumpff et al. showed [8] that tyrosine kinase inhibitors, like the Genestein with different dozes, reactivate the maxi-channels BKCa2+ sequestered by the tyrosine kinases. This relevant founding let understanding that repolarization

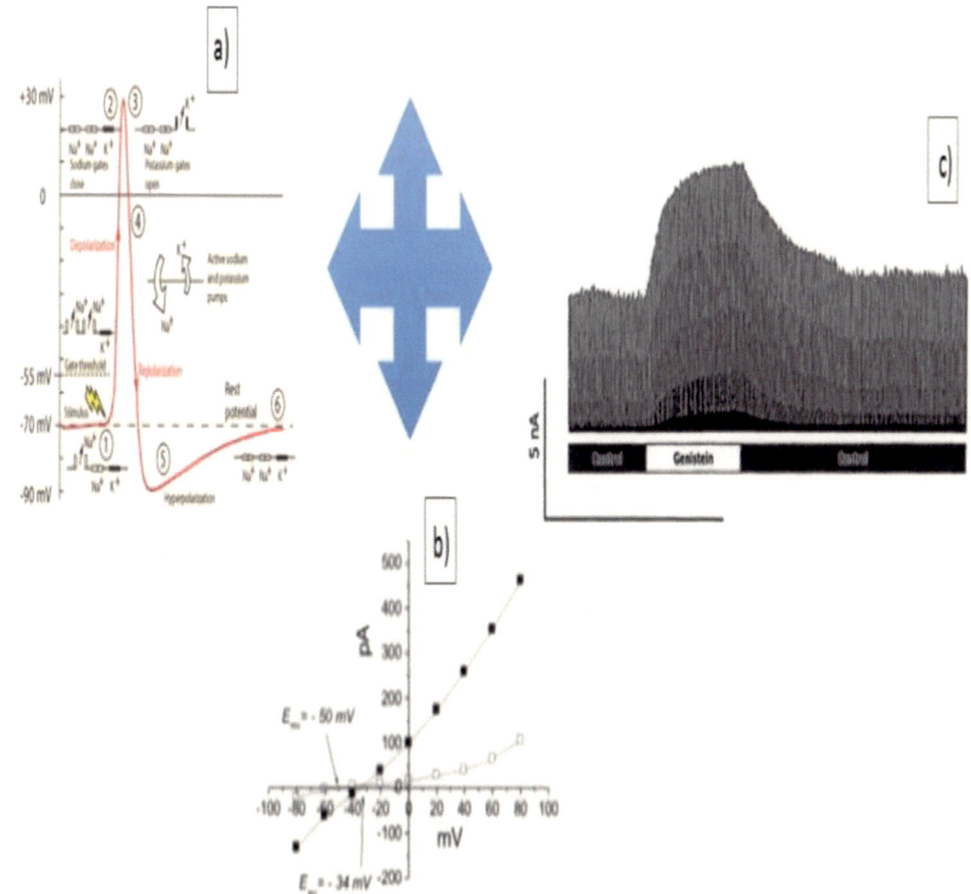

Figure 2.
*We illustrate in **Figure 2a** the typical TM cell membrane potential V_m, and its depolarization, repolarization and hyperpolarization regions. We notice that the regulation of ions flux through the cells (K, N_a, etc.) depends strongly on the right V_m levels. The dysfunction of the V_m potential keeping TM cells in the depolarization region, spouses high levels of water inside the cell due to lower KBC_a currents, this increases TM cell volume avoiding the output of the AH flux through TM cells. Then this V_m potential dysfunction increases intraocular pressure. **Figure 2b** shows V-I relationship between glaucoma GTM and healthy TM cells. **Figure 2c** shows the set of BKCa2+ ionic currents measured by Grant after the application of tyrosine kinase inhibitors*

and hyperpolarization region in TM cells play a central role in TM inflammatory dysfunction. We have concluded from this finding that relaxation of the trabecular meshwork is related with a dysfunction of the membrane voltage. Consequently, we can assure that POAG is a membrane potential V_m illness due to the dysfunction of BKCa2+ channels that affects the membrane potential V_m, keeping the cell membrane at the repolarization region for long time avoiding relaxing the TM cells.

A relevant result showed by Stuff [8] was that the re-activation of BKCa2 channels resulted with the application of an agonist of tyrosine kinase, showing that ionic current of BKCa2 are rectifier channels type [8–10]. Stumpff results let us understand that inhibitors of tyrosine kinase recovers the ionic current flow trough the channel, and that as a consequence of that V_m goes a to normal status. We infer that the inverse is feasible too, recovering the BKC_a+ open and close functionality by the right membrane potential. This is can be done from the application of an exogenous membrane $V_{m.}$ voltage, if it is properly selected. This principle is widely used in Electrophysiology where patch clamp techniques apply a set DC voltage from positive to negative values to stimulate the cell membrane to measure the answer of the ionic currents in the cell varying the membrane voltage. Based on the above considerations we formulate the hypothesis that to recover the functionality of the BKC_a+ maxi-channel can be achieved by a right exogenous membrane potential V_m. We proposed as our main working hypothesis that this can be achieved through an specific exogenous voltage applied to the tissue. A properly chosen exogenous voltage to the tissue will provide the right repolarization, depolarization and hyperpolarization voltages over the membrane of the TM cells in order to recover its flexibility.

Another important result reported by Grant, et al., [10] shows that the V-I relationship between glaucoma GTM and healthy TM cells is clearly different in Electrophysiology analysis, as shown in **Figure 2b**. They study the performance of BKC_a+ maxi-channel Ionic channels expressed in GTM and healthy TM cells, founding important differences among its current measured from a set of DC values. Reported results show important differences in the V-I relationship between GTM and TM for each specific ionic currents studied, consequently the performance regulation of normal TM is different from glaucoma GTM cells. Those differences founded in the electrophysiological properties of both groups let us to understand that POAG glaucoma illness is a dysfunction of the membrane potential of the GTM cells [9–12]. **Figure 2c** shows that for healthy TM cells BKCa+ is a rectifier current in the depolarization zone, however GTM shows a depressed BKCa+ current for positive DC voltages.

Daniel A. Ryskamp, et al. Published in 2016 an interesting paper about how TRPV4 regulates calcium homeostasis in mammalian eye. A. Ryskamp, et al. [12] shows that TRPV4 inhibition lowers IOP in GTM cells. They show too that an agonist of TRPV4 is a potential protector of the optic nerve. Due to TRPV4 are transient voltage dependent channels they could be manipulated, by exogenous voltages. TPES has demonstrated its beneficial effects in Retinosis Pigmentaria and IOP control properties [13–18], however new research is required to evaluate its protecting effects.

Figure 2b shows V-I relationship between glaucoma GTM and healthy TM cells. The voltage current V-I for healthy TM and GTM is clearly different. Healthy TM cells V-I curve shown in black squares is a typical rectifier function. However GTM in white squares vary its rest potential and decreases the V-I slope cells compared with healthy TM cells, this strongly modifies V_m. membrane potential in GTM cells. This V_m dysfunction keeps GTM cells in depolarization region, then avoiding the right control of the TM cells volume.

Figure 2c shows the set of BKCa2+ ionic currents measured by Grant after the application of tyrosine kinase inhibitors that goes back TM cells to normal V_m membrane potential. Signals in **Figure 2c** fits more alike rectifier function depending on the tyrosine kinase inhibitors dosage.

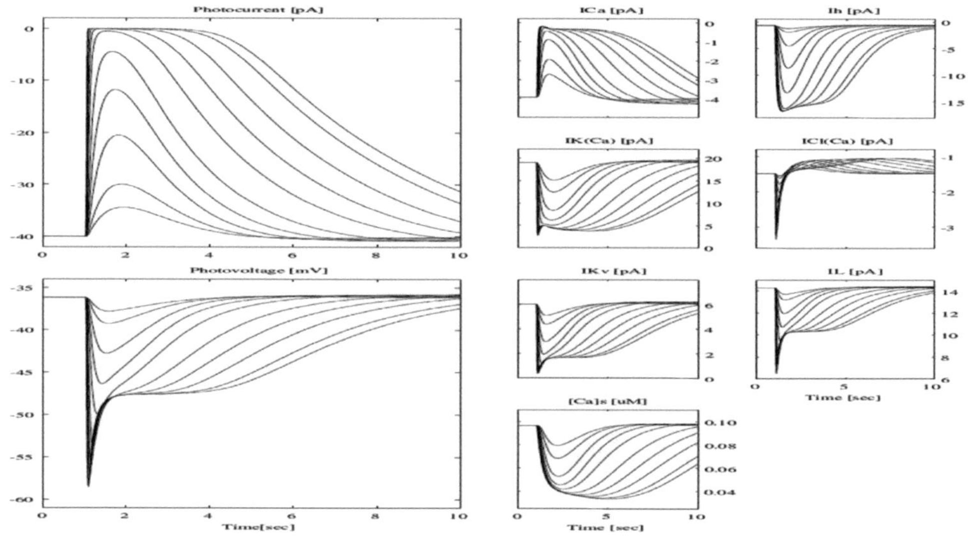

Figure 3.
Our stimulator system can reproduce any ionic current from a mathematical model or from any real biological ionic currents with 99.9% accuracy. Typical calcium-activated chloride current (ICl(Ca)) and calcium-activated potassium current (IK(Ca)) like Usui's model [11], and as many others ionic currents must be available in dedicated TPES stimulators.

2.1 Micro stimulation procedure

The set of signals that fits more alike BKCa2+ after the application of tyrosine kinase inhibitors is shown in **Figure 2c** according with Stuff et al reports [8]. Consequently an exogenous voltage can be used as an adjuvant treatment to control IOP in patients with POAG. The waveform shown in **Figure 2c** fits more accurately with BKCa2+ ionic cells in TMC, however an enormous set of different signals conform the ionic currents environment in any cell. We show in **Figure 3** a survey of the ionic currents that could be required to stimulate specific VDCH. A deeper understanding about the effects of TES in GTM cells is required to adjust the optimum parameters in TES, among others, waveform, frequency, amplitude and DC offset. TES requires special adjustment to fix the repolarization, depolarization and hyperpolarization zone for each target.

Our TES device reproduces electric profiles analogous to those reported by Stumpff [8, 9] in the BKCa2+ after the application of tyrosine kinase inhibitors, as well as many other ionic signals like Usui's mode [11]. The flexibility to generate any desired action potential opens fresh opportunities to brand new experiments in TPES. Our stimulator system can reproduce any real biological ionic currents with 99.9% accuracy. These new approaches to stimulate TMCs open brand new opportunities to understand more precisely the neuro-regulation effects of electrical stimulation in glaucoma and other degenerative illness.

2.2 Transpalpebral electrical microstimulation

TPES is not an invasive procedure as seen in **Figure 4**. This treatment modality is advantageous to conventional medical treatment consisting of hypotensive ocular drugs, which are associated with local and possible systemic side effects. This procedure induces no changes on the ocular surface in opposition to ocular surface changes induced by the topical medication and the preservatives. No inflammatory phenomenon is generated; moreover, the immunological apparatus remains unaltered.

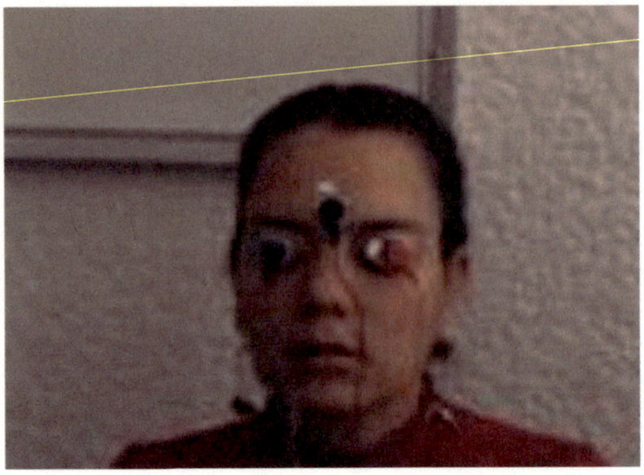

Figure 4.
Shows the Transpalpebral electrical stimulation procedure on a POAG patient. TPES is applied over the eyelid by an electrode array and through an electronic system with control of wave-forms and its parameters, like: Amplitude, frequency, DC offset and stimulation time.

Electrical micro stimulation focuses to the true cause of ocular hypertension, that is, the treatment of the dysfunctional trabeculum, while the rest of the therapy is focused on reducing the production of the aqueous humor or the exit of it through the unconventional route. In addition, it is important to emphasize the fact that this approach is highly cost efficient, when compared to anti-glaucoma therapy. We report here, a one year follow-up clinical study with patients with POAG.

2.3 A non-conventional signal generator

A non-conventional TPES system to stimulate POAG patients requires the developed of friendly electronic systems with strict biomedical standards to assure no damage to the eye, keeping very low current inside the ocular glove and no temperature effects over the eyelid. It requires too a software interface that let call any one of pre programmed ionic currents. A graphical user interface allows in our system to choose among sixteen different stimulation signals. The Hardware-software based waveform generator can produce practically any signals, each one characterized by its shape, amplitude, and frequency. Waveform shape is acquired from a set of discrete X(n) waveforms previously defined, the reader can find important discussion about ionic currents parameters and waveforms in [19–22]. Waveforms can be acquired also graphically by a virtual draw tool or even by a mathematical model, Usui in [11] shows an interesting model to display waveforms from the retina complex network, however we require more research to find the optimum waveform parameters to apply in clinic. The selected TES stimulator signals are transmitted using a simple USB port connected to microcontroller that generates the stimulation signal required. The stimulation system designed in our group is able to recreate any complex waveform as the ones reported in electrophysiology cell. This scheme can be used to obtain a class of multichannel stimulator that can be the core part of several biomedical applications. The desired action potential is selected by a friendly software showed in **Figure 5**.

A set of sixteen predefined waveforms, as shown in **Figure 3** let the physician select the signal and the parameters required, see **Figure 5**. The selected signal is programed into a microcontroller's memory, allowing converting the digital data into an analogue output signal. Then the output of an electronic system can be connected to the microelectrodes array to stimulate the cornea, as shown in **Figure 4**.

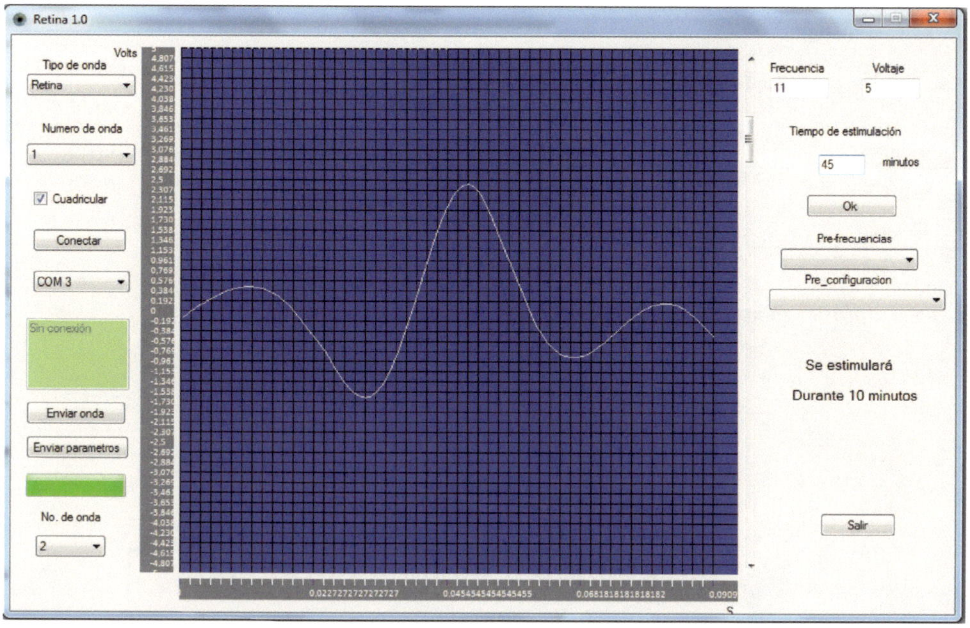

Figure 5.
Shows TPES interface system that let the physician to select the waveform among a set of sixteen pre defined V_m potentials to apply in TPES. The interface let define all parameters in TPES like: stimulation time, frequency, DC offset and voltage stimulation range.

3. Clinical founding

A year Follow-up of patients without surgical treatment (with and without topical treatment) is reported in this section. TPES is applied over the eyelid by an electrode array and through an electronic system with control of wave-forms and its parameters, like: amplitude, frequency, DC offset, stimulation time.

We report in this paper a one year TES study In 78 eyes from 46 patients (pilot group), 35 eyes of which 34.2% (n = 12) were specifically applied micro-stimulation and 65.8% (n = 23) applied micro-stimulation and topical treatment. The mean age was 65.67 years (44–80 years), 66.6% (n = 12) were women and 33.3% (n = 6) men. The mean glaucoma diagnosis time for our sample was 82.06 months ±58.6 (**Tables 1** and **2**).

3.1 Results

During the follow-up of the patients, the IOP decrease was recorded for one year, baseline IOP = 18.45 mmHg ± 2.45, one month later IOP = 15.85 mmHg ± 3.03, 3 months later IOP = 13.88 mmHg ± 1.90, after 6 months 14.65 mmHg ± 2.20 and

Variable	Man	Women	Total
Eyes (n)	12	23	35
Gender (%)	33.3	66.6	100
Mean Age ± DS (years)	68.33 ± 6.532	64.33 ± 9.717	65.8 ± 8.798
Rank (min,max)	63,80	44,77	44,80

Table 1.
Demographic data of the patients in the pilot group, one year follow-up.

Follow-up (months)	95% confidence interval			
	Mean	Inferior	Superior	Sig. (*bilateral*)
1	2.60000	1.78284	3.41716	< 0.001
3	4.57143	3.68859	5.45426	< 0.001
6	3.80000	2.82555	4.77445	< 0.001
12	3.80000	2.64317	4.95683	< 0.001

Table 2.
Intraocular pressure reduction mean in the pilot group, after one-year follow-up.

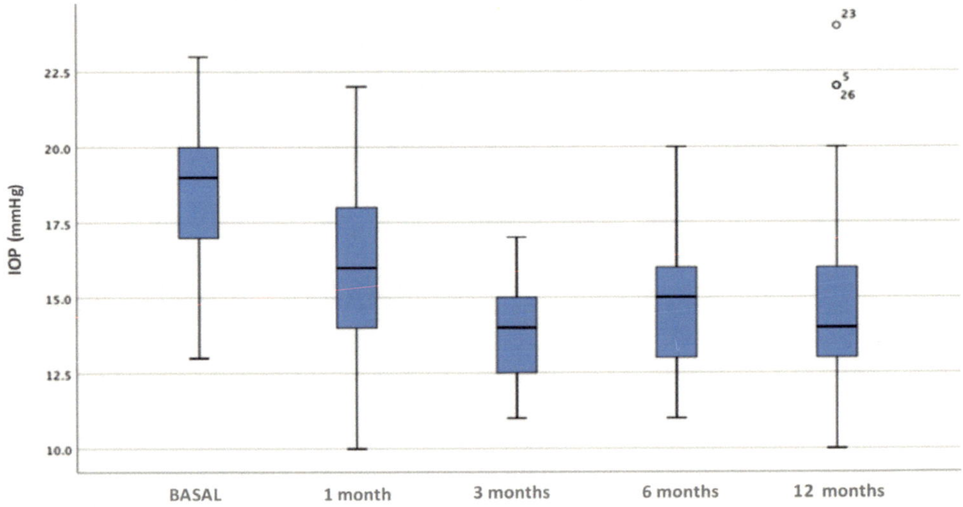

Figure 6.
Behavior of intraocular pressure in the one-year follow-up of the pilot group.

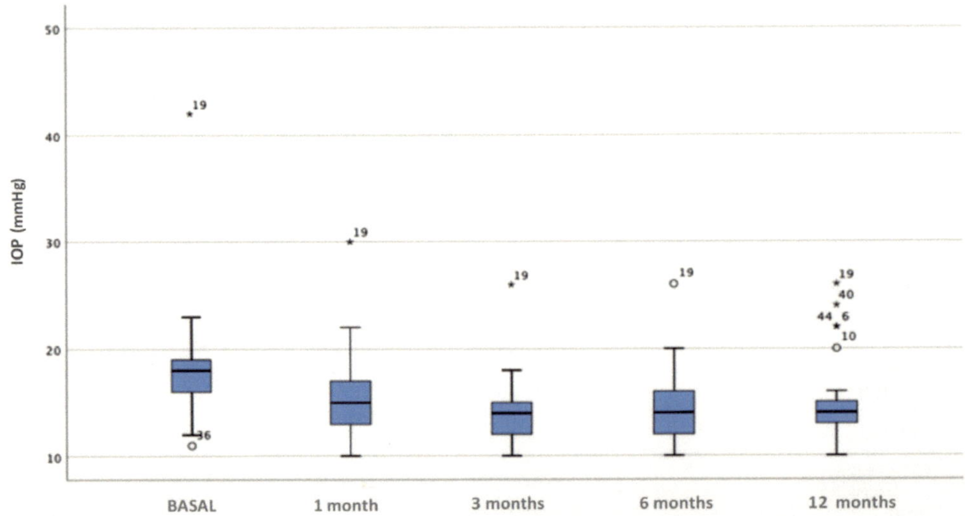

Figure 7.
Box diagram of intraocular pressure behavior at one-year follow-up of the group of patients with GPAA.

finally after one year IOP = 14.65 mmHg ± 3.09. Total decrease of 20.59% of the IOP after one year.

Total sample included 74 eyes of 40 patients, of which: 82.5% (n = 33) patients diagnosed with primary open-angle glaucoma, 2.5% (n = 1) with neovascular glaucoma, 2.5% (n = 1) with ocular hypertension and 12.5% (n = 5) with congenital glaucoma. The average age was 59.03 years (12–81 years), 57.5% (n = 23) were women and 42.5% (n = 17) men. The mean glaucoma diagnosis time for our sample was 125.43 months ±159.52 (**Figures 6–11**).

Reduction of intraocular pressure in patients with GPAA, dividing the sample into 3 groups. From the previous sample, patients with GPAA were selected. 65 eyes of 34 patients were included. The mean age was 62.53 years (35–81 years), 58.82%

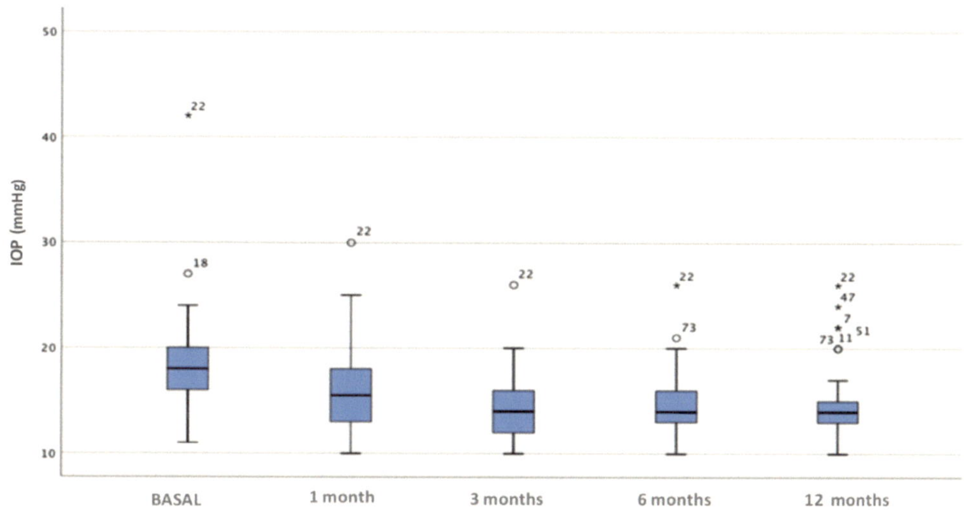

Figure 8.
Intraocular pressure behavior at one-year follow-up in the group of patients with GPAA, GN, GC and HTO.

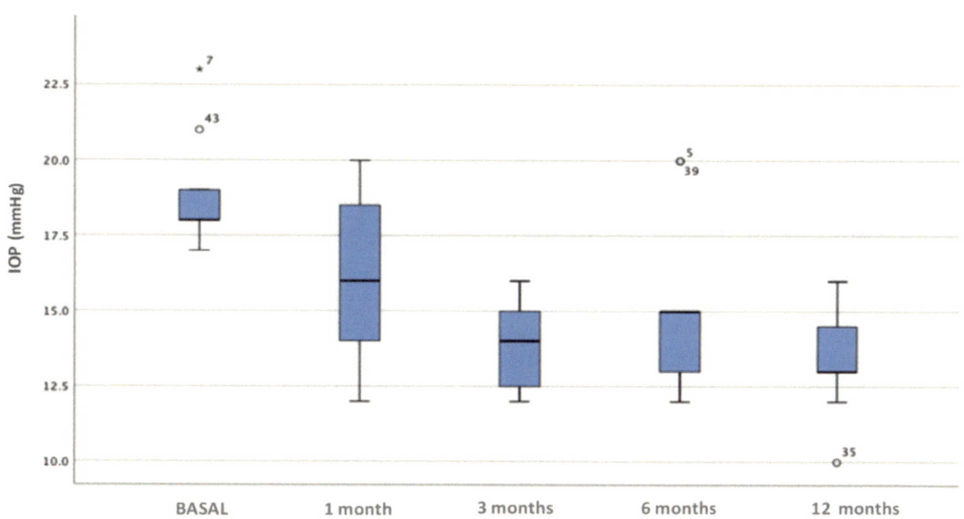

Figure 9.
Intraocular pressure behavior at one-year follow-up of group 1 [microstimulation only].

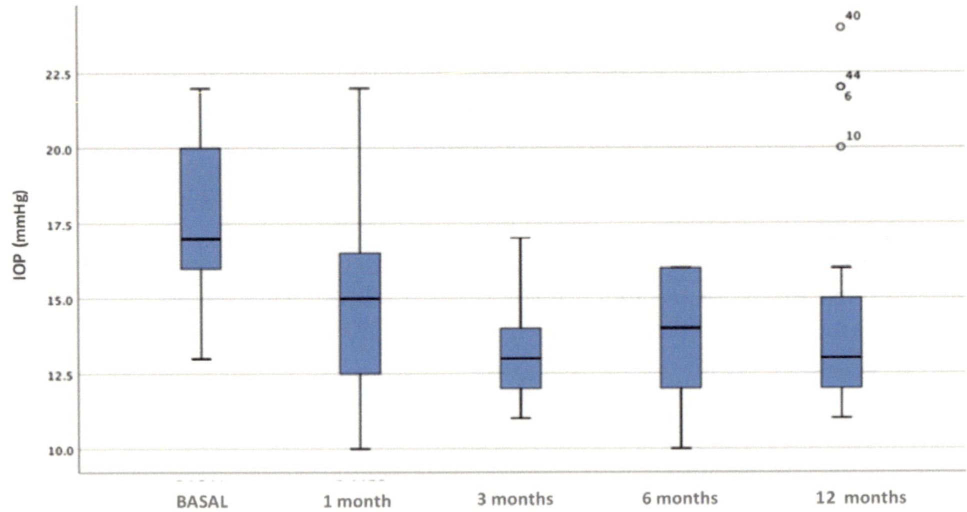

Figure 10.
Intraocular pressure behavior at one-year follow-up of group 2 [microstimulation + maximum tolerated drug treatment].

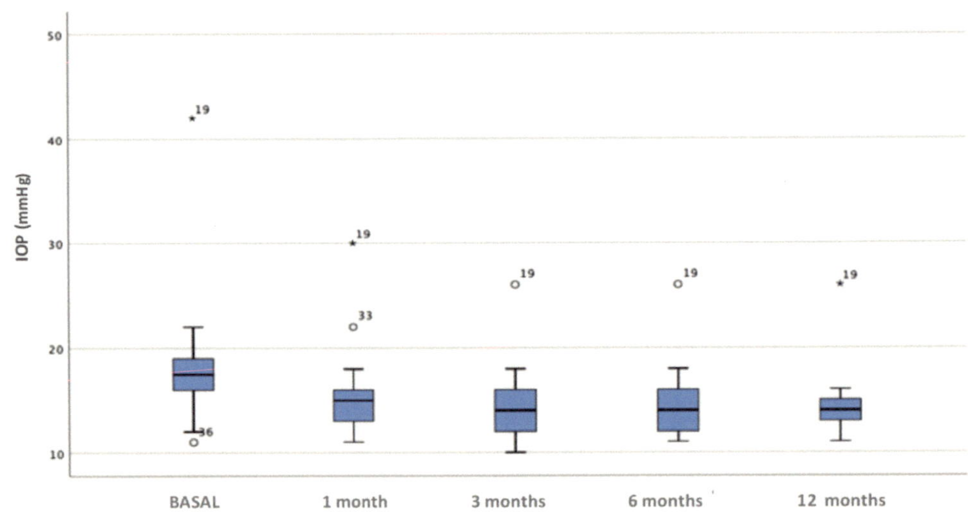

Figure 11.
Intraocular pressure behavior in one-year follow-up of group 2 [microstimulation + maximum tolerated pharmacological treatment + surgical treatment].

(n = 20) were women and 41.17% (n = 14) men. The mean time of diagnosis of glaucoma for our sample was 88.94 months ±69,289 (**Tables 3–5**).

The sample was divided into 3 groups according to the treatment received. Group 1 [microstimulation only] 17.64% (n = 6), group 2 [microstimulation + maximum tolerated drug treatment) 41.18% (n = 14), group 3 [microstimulation + pharmacological + surgical] 41.18% (n = 14). The average number of sessions per week was 1.71. In the total sample, the baseline IOP was 18.09 mmHg ± 3.97, 15.33 mmHg ± 3.37 a month, 13.86 mmHg ± 2.51 at 3 months, 14.21 mmHg ± 2.55, and 14.27 mmHg ± 2.91 at 6 months. With a decrease in IOP per year of 21.11% (**Table 6**).

Group 1 [microstimulation only] the baseline IOP was 18.83 mmHg ± 1.64, at a month of 16.16 mmHg ± 2.85, at 3 months 13.83 mmHg ± 1.46, at 6 months 15.00 mmHg ± 2.55 and at 13.41 mmHg ± 1.56. IOP decrease per year of 28.78% (**Table 7**).

Variable	Man	Women	Total
Eyes (n)	31	43	74
Gender (%)	41.9	58.1	100
Mean Age ± DS (years)	57.59 ± 17.256	60.09 ± 15.687	59.03 ± 16.203
Rank (min,max)	12,80	17,81	12,81

Table 3.
Demographic data of patients with GPAA, GN, GC and HTO.

Follow-up (months)	95% confidence interval			
	Mean	Inferior	Superior	Sig. (*bilateral*)
1	2.72973	2.21911	3.24035	< 0.001
3	4.22973	3.58742	4.87204	< 0.001
6	3.91892	3.22044	4.61740	< 0.001
12	3.98649	3.20199	4.77098	< 0.001

Table 4.
Mean intraocular pressure decrease in the one-year follow-up of the group with GPAA, GN, GC and HTO.

Variable	Man	Women	Total
Eyes (n)	27	38	65
Gender (%)	41.53	58.46	100
Mean Age ± DS (years)	60.64 ± 13.449	63.85 ± 11.018	62.53 ± 11.988
Rank (min,max)	36,80	35,81	35,81

Table 5.
Demographic date of the group of patients with GPAA.

Follow-up (months)	95% confidence interval			
	Mean	Inferior	Superior	Sig. (*bilateral*)
1	2.75385	2.17932	3.32837	< 0.001
3	4.23077	3.53436	4.92718	< 0.001
6	3.87692	3.14329	4.61055	< 0.001
12	3.81538	2.98672	4.64405	< 0.001

Table 6.
Average decrease in intraocular pressure in the one-year follow up of the group with GPAA.

Group 1 [microstimulation only] the baseline IOP was 18.83 mmHg ± 1.64, at a month of 16.16 mmHg ± 2.85, at 3 months 13.83 mmHg ± 1.46, at 6 months 15.00 mmHg ± 2.55 and at 13.41 mmHg ± 1.56. IOP decrease per year of 28.78% (**Table 8**).

Group 2 [microstimulation + maximum pharmacological treatment tolerated) the baseline IOP was 17.70 mmHg ± 2.64, a month of 14.81 mmHg ± 3.07, at 3 months 13.29 mmHg ± 1.95, at 6 months 13.74 mmHg ± 1.99 and at 14.48 mmHg ± 3.50. With a decrease in IOP per year of 18.19%. It should be noted that 4 of the 14 patients were able to suspend 1 to 3 medications during the follow-up year.

Follow-up (months)	95% confidence interval			
	Mean	**Inferior**	**Superior**	**Sig. (*bilateral*)**
1	2.66667	.80274	4.53059	.009
3	5.00000	3.98630	6.01370	< 0.001
6	3.83333	1.77302	5.89365	.002
12	5.41667	4.04969	6.78364	< 0.001

Table 7.
Average decrease in intraocular pressure in the one-year follow-up of group 1 [microstimulation only].

Follow-up (months)	95% confidence interval			
	Mean	**Inferior**	**Superior**	**Sig. (*bilateral*)**
1	2.88889	2.07275	3.70503	< 0.001
3	4.40741	3.29792	5.51689	< 0.001
6	3.96296	2.93092	4.99500	< 0.001
12	3.22222	1.88147	4.56297	< 0.001

Table 8.
Mean decrease in intraocular pressure in the one-year follow-up of group 2 [microstimulation + maximum pharmacological treatment].

Follow-up (months)	95% confidence interval			
	Mean	**Inferior**	**Superior**	**Sig. (*bilateral*)**
1	2.65385	1.70527	3.60242	< 0.001
3	3.69231	2.39592	4.98869	< 0.001
6	3.80769	2.48514	5.13024	< 0.001
12	3.69231	2.23479	5.14983	< 0.001

Table 9.
Mean decrease in intraocular pressure in the one-year follow-up of group 3 [microstimulation + maximum pharmacological treatment + surgical treatment].

Group 3 [microstimulation + maximum tolerated pharmacological treatment + surgery] the baseline IOP was 18.15 mmHg ± 5.61, at a month of 15.50 mmHg ± 3.8, at 3 months 14.46 mmHg ± 3.24, at 6 months 14.34 mmHg ± 3.03 and per year 14.46 mmHg ± 2.73. With a decrease in IOP per year of 20.33%. It should be noted that 3 of the 14 patients were able to suspend 1 to 2 medications during the year of follow-up (**Table 9**).

4. Conclusions, future and challenges

Transpalpebral Electrical Stimulation focuses to the true cause of ocular hypertension, that is, the treatment of TM dysfunction, while the rest of the therapy is focused on reducing the production of the aqueous humor or the exit of it through the unconventional route [19, 20]. TPES is not only an alternative IOP therapy but also is a new way to face degenerative diseases, since the pathology is treated from another vision; the dysfunction of the voltage dependent channels looking its functionalization from an electronic medication. Electric medication is not new, however is not understood at all. The physiopathology bases discussed in this paper opens new alternatives to deal other degenerative illness where dysfunction of VDCH can be identified as the cause of the illness, some of them as an inflammatory cell process.

TPES requires to study the effects of electrical stimulation in both: cells, tissues and complex structures like the eyeball. This new approach requires the development of new flexible biomedical tools to stimulate the eyeball. A new future is coming with Biochips inside the eye to electrically stimulate the cornea. New challenges technology development requires TES in both: clinical applications and in cell culture to study the effects of electric stimulation of the trabecular mesh and some new alternatives to couple electrodes with both electronic systems and eye ball tissue. New challenges faces the need to measure inside the eyeball IOP integrated with biochips to stimulate the trabecular mesh to close the challenge to measure simultaneously IOP and control it by electric stimulation. New biochips with flexible substrates like polyamide emerge as a new biocompatible material to be integrated inside the eye to build complex systems to help people with glaucoma [23].

We propose a new medication procedure through the application of waveforms that can be generated from electronic technology feasible at hand. These new alternatives to treat degenerative ocular illness face new challenges and opportunities. However there is a long way to understand more precisely the neuro-protection effects of electrical stimulation at the trabecular mesh and at the inner retina. Most of diseases derivative in vision lost are related to Open Angle Glaucoma, Retinal degenerations (RDs), Retinitis pigmentosa (RP) and Age-related macular degeneration (AMD). It seems to be that most of them are related with a dysfunction of several voltage dependent ionic channels.

Optimal TPES parameters to target specific VDCH is still a challenge. New experiments are required to evaluate the effects of firing specific inner retinal neural structures. In our opinion, we show in this paper excellent results in controlling IOP during one year TPES stimulation as shown above. This pushes to look for new solutions to degenerative ocular illness.

Acknowledgements

The present work was supported by a grant from the CONACYT (Grant no. 0261660), and The Instituto Politécnico Nacional de México, grant No. 20196221.

Conflicts of interest

The authors declare that they have no conflicts of interest.

Author details

Luis Nino-de-Rivera[1*], Diego Cervera[2] and Paola Castillo-Juarez[3]

1 ESIME UC, Artificial Vision Lab, Instituto Politécnico Nacional, IPN, México

2 APEC, Hospital "Luis Sánchez Bulnes", México

3 ENCB, Virology Lab, Instituto Politécnico Nacional, IPN, México

*Address all correspondence to: luisninoderivera@gmail.com

References

[1] Tham YC, Li X, Wong TY, Quigley HA, Aung T, Cheng CY. Global prevalence of glaucoma and projections of glaucoma burden through 2040: A systematic review and meta-analysis. Ophthalmology. 2014;121:2081-2090.

[2] Schmidl D, Schmetterer L, Garhöfer G, Popa-Cherecheanu A. Pharmacotherapy of glaucoma. J Ocul Pharmacol Ther. 2015;31:63-77.

[3] Saccà SC, Pulliero A, Izzotti A. The dysfunction of the trabecular meshwork during glaucomacourse. J Cell Physiol. 2015;230:510-525.

[4] Anderson DR, Drance SM, Schulzer M, Comparison of glaucomatous progression between untreated patients with normal-tension glaucoma and patients with therapeutically reduced intraocular pressures. Am J Ophthalmol 1998;126:487-497.

[5] Tsai JC. A comprehensive perspective on patient adherence to topical glaucoma therapy. Ophthalmology. 2009;116:30-36.

[6] Robin AL, Covert D. Does adjunctive glaucoma therapy affect adherence to the initial primary therapy Ophthalmology. 2005;112:863-868.

[7] Stamer WD, Acott TS. Current understanding of conventional outflow dysfunction in glaucoma. Curr Opin Ophthalmol. 2012;23:135-143.

[8] Stumpff F, Que Y, Boxberger M, Strauss O, Wiederholt M, Stimulation of maxi-K channels in trabecular meshwork by tyrosine kinase inhibitors, Invest Ophthalmol Vis Sci.1999;40:1404-1417.

[9] Soto D, Comes N, Ferrer, et al. Modulation of aqueous humor outflow by ionic mechanisms involved in trabecular meshwork cell volume regulation. Invest Ophthalmol Vis Sci. 2004;45:3650-3656.

[10] Grant J, Tran V, et al. Ionic currents of human trabecular meswork cells from control and glaucoma subjects. J Membr Biol. 2013;246: 167-175.

[11] Kamiyama, Y., Wu, S.M. and Usui, S. (2009) Simulation analysis of bandpass filtering properties of a rod photoreceptor network. Vision Research, 49, 970-978.http://dx.doi.org/10.1016/j.visres.2009.03.003

[12] A. Ryskamp, TRPV4 regulates calcium homeostasis, cytoskeletal remodelling, conventional outflow and intraocular pressure in the mammalian eye. Scientific Reports | 6:30583 | DOI: 10.1038/srep30583. www.nature.com/scientificreports/

[13] Felix Gil et al, Transpalpebral Electrical Stimulation as a Novel Therapeutic Approach to Decrease Intraocular Pressure for Open-Angle Glaucoma: A Pilot Study Hindawi Journal of Ophthalmology Volume 2018, Article ID 2930519, 6 pages doi:10.1155/2018/2930519

[14] Daniel Robles-Camarillo, Luis Niño-de-Rivera, Jessica López-Miranda, Félix Gil Carrasco, Hugo Quiroz-Mercado The effect of transcorneal electrical stimulation in visual acuity: Retinitis pigmentosa J. Biomedical Science and Engineering, 2013, 6, 1-7. doi:10.4236/jbise.2013.610A2001 Published Online October 2013 (http://www.scirp.org/journal/jbise/).

[15] Schatz, A., et al. (2011) Transcorneal electrical stimulation for patients with retinitis Pigmentosa: "A prospective, randomized, sham-controlled exploratory study". Investigative ophthalmology and visual science, 52.

[16] Gelkeler, F., et al. (2006) Phosphenes electrically evoked with DTL electrodes: A study in patients with retinitis Pigmentosa, glaucoma, and homonymous visual field loss and Normal subjects. Investigative Ophthalmology and Visual Science, 47, 4966-4974. doi:10.1167/iovs.06-0459

[17] Schatz, A., et al. (2011) Transcorneal electrical stimulation for patients with retinitis Pigmentosa: "A prospective, randomized, sham-controlled exploratory study". Investigative Ophthalmology & Visual Science, 52.

[18] Naycheva, L., et al. (2013) Transcorneal electrical stimulation in patients with retinal artery occlusion: A prospective, randomized, sham-controlled pilot study. OphthalmolTher, 2, 25-39.http://dx.doi.org/10.1007/s40123-013-0012-5.

[19] Morimoto, T., et al. (2010) Optimal parameters of Trancorneal electrical stimulation (TES) to Be neuroprotective of Axotomized RGCs in adult rats. Experimental Aye Research, 90, 285-291.

[20] Xie, et al. (2011) Modeling and percept of Transcorneal electrical stimulation in humans. IEEE Transactions on Biomedical Engineering, 58, 1932-1939.

[21] Ackerman, M.J., and Clapham, D.E. (1997). Ion channels-basic science and clinical disease. N. Engl. J. Med. 336, 1575-1586.

[22] Miramoto, T., et al. (2005) Transcorneal electrical stimulation rescues Axotomized retinal ganglion cells by activating endogenous retinal IGF-1 system. Investigative Ophthalmology & Visual Science, 46, 2147-2155.

[23] Félix Gil Carrasco, Daniela Díaz Alonso, Luis Niño-de-Rivera, "Biocompatibility and implant of a less invasive intraocular pressure sensor". Microelectronic Engineering 159 (2016) 32-37, http://dx.doi.org/10.1016/j.mee.2016.02.004

Chapter 7

Glaucoma Related to Ocular and Orbital Tumors

Sonal P. Yadav

Abstract

Secondary glaucoma due to ocular and orbital tumors can be a diagnostic challenge. It is an essential differential to consider in eyes with a known tumor as well as with unilateral, atypical, asymmetrical, or refractory glaucoma. Various intraocular neoplasms including iris and ciliary body tumors (melanoma, metastasis, lymphoma), choroidal tumors (melanoma, metastasis), vitreo-retinal tumors (retinoblastoma, medulloepithelioma, vitreoretinal lymphoma) and orbital tumors (extra-scleral extension of choroidal melanoma or retinoblastoma, primary orbital tumors) etc. can lead to raised intraocular pressure. The mechanisms for glaucoma include direct (tumor invasion or infiltration related outflow obstruction, trabecular meshwork seeding) or indirect (angle closure from neovascularization or anterior displacement or compression of iris) or elevated episcleral venous pressure secondary to orbital tumors. These forms of glaucoma need unique diagnostic techniques and customized treatment considerations as they often pose therapeutic dilemmas. This chapter will review and discuss the mechanisms, clinical presentations and management of glaucoma related to ocular and orbital tumors.

Keywords: ocular tumors, secondary glaucoma, orbital tumors, angle infiltration, neovascular glaucoma, neoplastic glaucoma

1. Introduction

With the advent of constantly evolving and advancing ophthalmic imaging techniques as well as surgical modalities in the field of ophthalmic diseases, diagnostic accuracy, and treatment outcomes of ocular as well as orbital tumors have improved remarkably over the past few years. Raised intraocular pressure (IOP) is known to be one of the presenting features or associated finding for numerous ocular as well as orbital tumors. Ocular and orbital tumors can cause secondary glaucoma due to various mechanisms. They often pose a diagnostic challenge as well as a therapeutic dilemma owing to the complex pathophysiology involved. A thorough clinical evaluation, appropriate index of suspicion and optimum use of ancillary testing can lead to a proper diagnosis and management in such scenario.

Intraocular tumors that can lead to secondary glaucoma are malignancies like iris melanoma, iris metastasis, iris lymphoma, ciliochoroidal melanoma, retinoblastoma as well as benign pathologies like iris melanocytoma, benign ciliary body medulloepithelioma, diffuse choroidal hemangioma. [1] Indicators for a possible underlying intraocular tumor are, markedly elevated and often asymmetric level of IOP, acquired iris heterochromia, glaucoma non-responsive to optimum treatment or accompanying distinctive ocular features. [1, 2] The mechanism by which these

IntechOpen

tumors can cause secondary glaucoma varies with tumor type, size, and extent of the main tumor as well as seeding, tumor location, growth pattern, ongoing treatment along with secondary features related to the tumor. [3] Iris and predominantly ciliary body tumors located in anterior segment can cause glaucoma by direct infiltration of anterior chamber angle or because of iris neovascularization; while large tumors originating in retina or choroid are likely to cause glaucoma following iris neovascularization because of long standing or total retinal detachment or secondary angle closure. [1, 2]

Orbital tumors which can be congenital, traumatic, inflammatory, vascular, or neoplastic in origin may cause secondary glaucoma due to mass effect or anatomical and vascular changes leading to raised IOP [4] Orbital tumors causing raised orbital pressure may directly increase the IOP by increasing hydrostatic pressure around the globe or indirectly by raising the episcleral venous pressure.

Management of secondary glaucoma due to ocular and orbital tumors depends on both tumor characteristics and glaucoma related factors. Treatment of primary tumor may lead to IOP control in some cases while for others, management options include medical management, laser trabeculoplasty, transscleral cyclophotocoagulation, anti-VEGF injections, minimally invasive glaucoma surgery (MIGS), filtering or shunting surgery or enucleation. Glaucoma surgery like filtering or shunting procedure can be performed with due caution in proven benign or completely regressed tumors post-treatment. Such surgeries in an eye with suspected but unproven benign/malignant ocular tumor must be avoided to prevent unintended iatrogenic tumor dispersion or seeding especially in cases with iridociliary tumors or retinoblastoma.

2. Mechanisms of glaucoma secondary to ocular and orbital tumors

A. Direct mechanism-

a. Solid tumor invasion- related outflow obstruction

b. Infiltrative tumor -related outflow obstruction

c. Trabecular meshwork seeding

B. Indirect mechanism-

a. Angle closure from neovascularization

b. Angle closure (compressive and rotational)

c. Ghost cell- Hemolytic

d. Elevated episcleral venous pressure

3. Ocular tumor related glaucoma

3.1 Anterior segment tumors

Anterior uveal tumors known to cause secondary glaucoma are iris/ciliary body melanocytoma, iris melanoma (nodular or diffuse), ciliary body melanoma (nodular or ring melanoma), iris lymphoma and iris metastasis. Direct invasion of the anterior chamber angle by infiltration followed by neovascularization and trabecular meshwork tumor seeding are the common etiologies of raised IOP in iris tumors; while pigment dispersion followed by direct angle invasion are the common etiologies for same in pigmented ciliary body tumors. [1, 2, 5, 6]

Iris melanocytoma, a variant of melanocytic iris nevus is a deeply pigmented benign tumor which is well circumscribed, often dark brown to black dome shaped lesion with cobblestone surface and feathery edges showing echogenic nodular thickening of iris on ultrasound biomicroscopy (UBM). (**Figure 1A–C**) [2, 7] Melanocytoma can undergo spontaneous necrosis with pigment dispersion leading to pigment-laden melanophages seeding the angle causing secondary glaucoma. [8] (**Figure 1D**) Secondary glaucoma has been reported in 11% of cases in a series of 47 iris melanocytoma where pigmented keratic precipitates and anterior chamber inflammation were identified as factors predictive of development of raised IOP emphasizing the role of macrophages in the anterior chamber angle. [7]

Suspected melanocytoma can be observed cautiously. They are very rarely known to show malignant transformation. [9] Clear corneal approach fine needle aspiration biopsy, minimally invasive Finger Iridectomy technique (FIT), iridectomy or iridogoniocyclectomy can be utilized to obtain histopathological diagnosis in atypical iris nevi and suspected melanocytoma. [7, 10] Secondary glaucoma demonstrating melanocytoma eyes can be treated medically, by transscleral photocoagulation, by sector iridectomy or with glaucoma filtration surgery. [7, 8, 11, 12] However, a diagnostic confirmation of the lesion by prior biopsy and histopathology

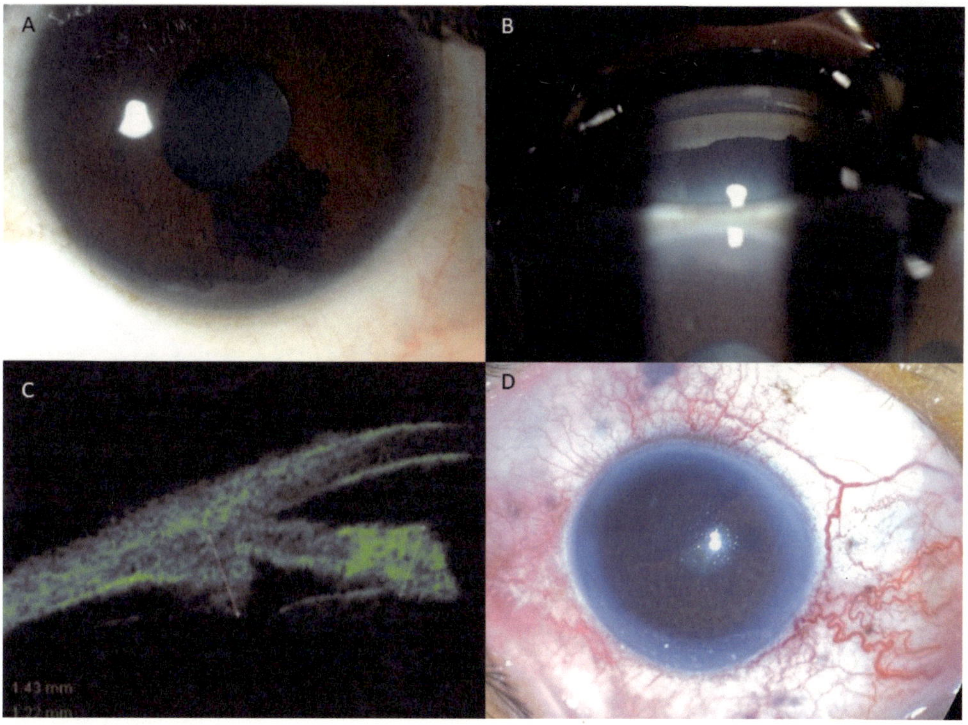

Figure 1.
(A) Slit lamp image of iris melanocytoma; (B) Gonioscopy image of iris melanocytoma with extension of pigmentation up to the angle, (C) ultrasound biomicroscopy (UBM) image of iris melanocytoma appearing as irregular hyperreflective lesion on iris surface; (D) large iris melanocytoma leading to raised IOP due to angle closure glaucoma (compressive angle closure) {image credit for **Figure 1A–C** *Dr. Paul T Finger MD, FACS, Director, The New York Eye Cancer Center, New York USA}.*

is mandated before planning a filtering surgery. Local surgical resection can be used to treat the secondary glaucoma caused by necrotic iris melanocytoma. [8, 13] Enucleation is reserved for painful blind eyes or eyes with absolute glaucoma.

Iris and iridociliary melanoma are the least common variant of uveal melanoma constituting only 2–3% of cases and have documented metastatic rate of 10.7–11%. [14–16]. Iris melanoma presents as a variably pigmented lesion (melanotic or amelanotic) which can be well circumscribed- nodular type, flat pigmented on iris surface- diffuse type and rarer predominantly trabecular meshwork involving type. (**Figure 2**) Secondary glaucoma has been reported in 33% of cases with iris melanoma while, with 100% of those with trabecular meshwork melanoma. [6, 17] Mechanisms include direct angle invasion (infiltration), iris neovascularization and trabecular meshwork seeding (melanomalytic glaucoma). [2, 3, 6] Unilateral findings of markedly elevated IOP, heterochromia iridis along with pigment dispersion onto the corneal endothelium, anterior iris surface and into the angle point towards melanomalytic glaucoma. [2, 18] Diffuse iris melanoma should be suspected in all cases of iris heterochromia and unilateral glaucoma. (**Figure 2B** and **D**) [2, 19] Morphological features of iris melanoma such as diffuse tumor location (>1 quadrant), posterior tumor margin involving the angle, reduced median tumor thickness (flat configuration) and greater extent of tumor seeding on the iris stroma and angle have been statistically related to presence of glaucoma at presentation. [6] Patients with iris melanoma and secondary glaucoma are at significantly higher risk of systemic metastasis with a hazard ratio of 4.51 compared to iris melanoma without

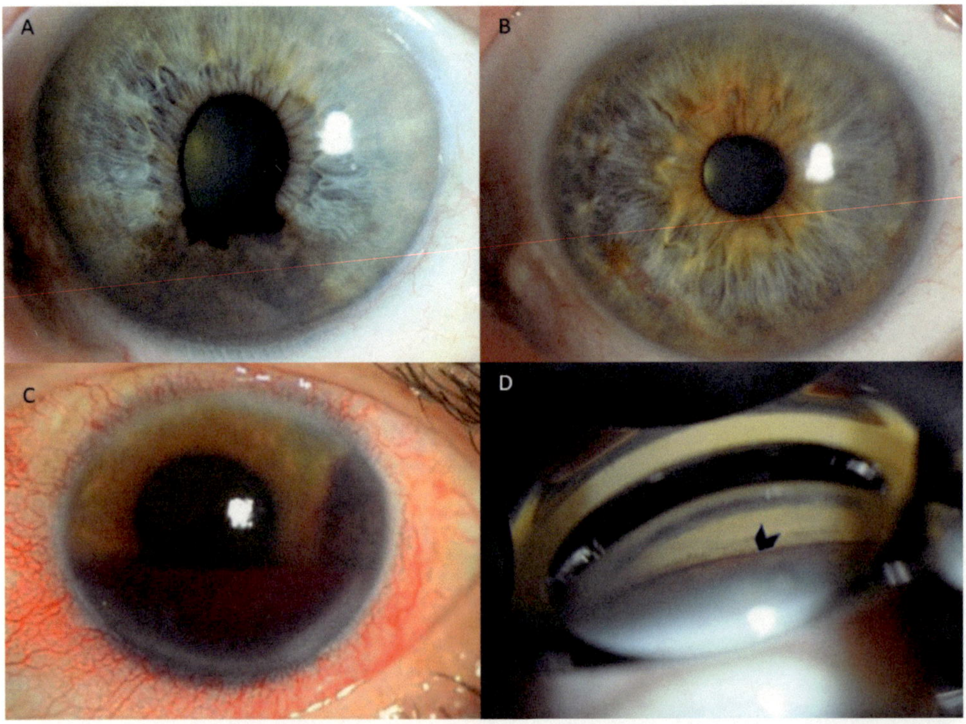

Figure 2.
(A) Slit lamp image of iris melanoma showing correctopia, ectropion uveae and extension up to the angle, (B) slit lamp image of diffuse iris melanoma presenting with angle closure, (C) slit lamp image of ciliary body melanoma with iris infiltration presenting as secondary glaucoma with hyphema. (D) Gonioscopy image of blood vessel (black arrowhead) at the anterior chamber angle in a case of diffuse iris melanoma with angle closure depicted in **Figure 2B**. *{image credit for* **Figure 1** *A, B, D- Dr. Paul T Finger MD, FACS, Director, The New York Eye Cancer Center, New York USA and for* **Figure 2C**- *Dr. Santosh G Honavar MD FACS, Director, Ocular Oncology Services, Centre For Sight Hospital, Hyderabad, India}.*

glaucoma. [3, 6] Possible explanation being-tumor location and its discohesive nature along with raised IOP that enables the egress of tumor cells into emissary veins leading to distant metastasis. [2, 6, 20].

Management of iris and iridociliary melanoma depends upon tumor size, location or extent, tumor seeding and presence of tumor related glaucoma. [15, 20] Local resection (iridectomy, iridocyclectomy), plaque brachytherapy, proton beam radiotherapy and enucleation are available treatment options for both nodular as well as diffuse iris melanoma. [6, 15, 20, 21] Secondary glaucoma in association with iris melanoma can be managed with medications, transscleral photocoagulation or laser trabeculoplasty. [22] Antivascular endothelial growth factor (anti-VEGF) injections can be tried in neovascular glaucoma. [23] Cases with refractory glaucoma resulting in blind painful eyes may warrant enucleation. [24] Filtering, shunting surgery or MIGS should be avoided in eyes with untreated tumor to prevent tumor spread outside of the globe. Diffuse iris melanoma or trabecular meshwork melanoma if misdiagnosed or missed prior to performing above named procedures can warrant enucleation for further tumor control. [3, 19]

Ring melanoma of the iris and ciliary body are a rare entity constituting only 0.3% of all uveal melanomas while having a poor prognosis owing to metastatic rate of up to 50%. [25]. The later could be likely due to delayed diagnosis as ring melanoma tend to grow circumferentially involving iris and entire ciliary body making them less obvious on ophthalmic examination. Prominent episcleral (sentinel) vessels, anterior chamber shallowing, multinodular tumor configuration, unilateral lens changes and occasional extra-scleral extension, light blockage on transillumination and ultrasonographic hollowness with intrinsic vascular pulsations are salient diagnostic features suggestive of ring melanoma. [25, 26] They tend to initially manifest with low intraocular pressure and later develop secondary glaucoma and secondary retinal detachment. Secondary glaucoma has been reported in 35% of cases. [25] The mechanisms involved are direct angle infiltration, trabecular meshwork seeding, angle closure and neovascularization of iris. [2, 3, 27–29] Plaque radiotherapy, proton beam radiotherapy or enucleation are available management options for the tumor. [25, 26, 30] Medical management, transscleral cyclophotocoagulation to uninvolved ciliary body or laser trabeculoplasty to uninvolved angle can be tried to control glaucoma in salvaged eyes. However, refractory cases require enucleation. [3] High index of suspicion is warranted when examining a case of unilateral refractory glaucoma with multinodular thickening of angle structures to avoid misdiagnosis as uveal effusion. Ultrasound biomicroscopy in such a case can help in identifying the tumor. Hurried shunting or filtering surgery for IOP control in tumor containing eyes can lead to tumor spread and further necessitate enucleation. [29, 31]

Iris and ciliary body metastasis constitute 8% and 2% of all cases of uveal metastasis respectively with breast and lung cancers being the most common primary tumors. [32] They tend to appear as solitary or multiple yellow, white, or pink stromal nodules with hyphema or pseudo hypopyon. They may present as ill-defined iris or ciliary body thickening in a setting of iridocyclitis. Ultrasound biomicroscopy (UBM) is useful in suspicious cases. (**Figure 3A** and **B**) Secondary glaucoma has been reported in about 37% of iris metastasis with angle invasion and iris neovascularization as mechanisms for raised IOP. [33] Treatment of primary, plaque brachytherapy, external beam radiotherapy (EBRT) or systemic chemotherapy are management options for iris and ciliary body metastasis. [2, 3, 33] Medical management, transscleral laser photocoagulation or laser trabeculoplasty can be useful for IOP control in these eyes. Anti VEGF injections may help in neovascular glaucoma. [3, 33–35]

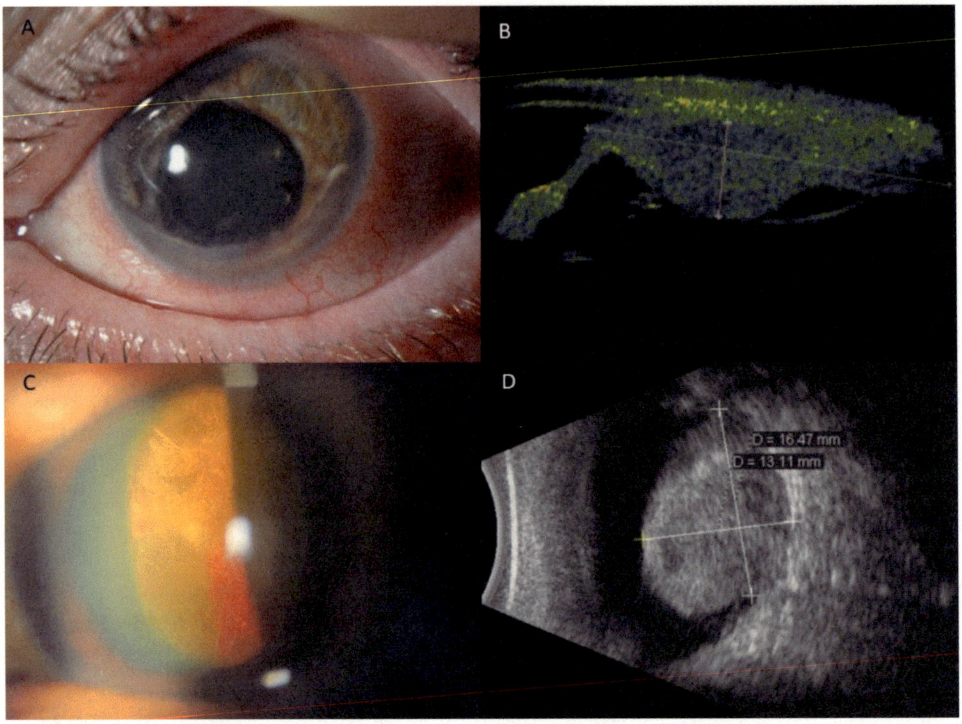

Figure 3.
(A) Slit lamp image of iris and ciliary body metastasis from lung adenocarcinoma showing correctopia, ectropion uveae, neovascularization of iris presenting with raised IOP (B) Ultrasound biomicroscopy image of iris metastasis with blunting of anterior chamber angle and thickened ciliary body (C) Slit lamp image of large choroidal melanoma presenting with intratumoral hemorrhage and raised IOP. (D) Ultrasound B scan of lesion in (C) showing low to moderately reflective choroidal melanoma with surrounding detached retina {image credit for **Figure 1** *A, B- Dr. Paul T Finger MD, FACS, Director, The New York Eye Cancer Center, New York USA}.*

Iris lymphoma, localized or diffuse is a differential diagnosis of iris melanoma or metastasis. Unilateral or bilateral presentation of diffuse non-granulomatous uveitis with raised IOP in open angles, uveitis- glaucoma- hyphema syndrome (UGH) or steroid resistant pseudo-uveitis are documented clinical manifestations of intraocular lymphoma. [36–38] The mechanism for raised IOP include angle infiltration, angle closure, hyphema. [3, 37] Plaque radiotherapy, EBRT, systemic chemotherapy have been used for management of iris lymphoma while, the raised IOP may respond to medical therapy, transscleral cyclophotocoagulation or laser trabeculoplasty. Enucleation is reserved for refractory cases. [2, 3]

Leukemic infiltration from acute lymphocytic leukemia (ALL) and acute myeloid leukemia (AML) in the anterior chamber, trabecular meshwork and Schlemm's canal can cause outflow obstruction and raised IOP. [2] ALL is the most common cause of both intraocular leukemia and secondary leukemia related glaucoma. Leukemia with central nervous system (CNS) involvement has been strongly associated with leukemic hypopyon and glaucoma. [2, 39] Treatment with systemic chemotherapy or EBRT is recommended for these. EBRT is known to show rapid resolution of angle infiltration in cases with secondary glaucoma. [2]

Other disorders like: Multiple myeloma, a malignant proliferation of plasma cells characterized by monoclonal production od immunoglobulins; Juvenile Xanthogranuloma, a benign histiocytic skin disorder can also affect anterior segment and lead to raised IOP. [2]

3.2 Posterior segment tumors

3.2.1 Choroidal tumors

Choroidal melanoma is the most common primary intraocular malignancy found in adults and the most common uveal melanoma encountered overall. Clinically it manifests as a dome shaped or mushroom shaped pigmented (melanotic) or non-pigmented (amelanotic) choroidal mass often associated with surrounding sub-retinal fluid, overlying orange pigmentation and exudative retinal detachment. (**Figure 3C**) The ultrasound B scan shows low to moderately reflective choroidal lesion with associated detached retina or occasional vitreous hemorrhage. (**Figure 3D**) Compared to anterior uveal melanomas, choroidal melanoma is less likely to cause secondary glaucoma (2%). [2, 3] Secondary glaucoma has been reported in both known and previously undiagnosed cases of choroidal melanoma. [2, 3] Secondary glaucoma has also been termed as the 'masquerading sign' for uveal melanomas in the literature. [40, 41] The reported mechanisms for raised IOP include iris neovascularization, direct angle invasion, angle closure, hyphema and suprachoroidal hemorrhage. [1, 3] Globe sparing episcleral plaque brachytherapy is commonly used modality of treatment followed by proton beam radiotherapy and local resection. Enucleation is warranted for large melanomas not amenable to radiotherapy or those causing absolute glaucoma. For control of IOP, medical management, transscleral cyclophotocoagulation or laser trabeculoplasty can be tried. In contrast to anterior uveal melanomas, for choroidal melanoma not involving ciliary body or iris root MIGS, filtering and shunting surgery can be considered in eyes where complete tumor regression has been achieved. [3]

Contrary to traditional thinking, a recent study of analysis of glaucoma drainage device surgery for control of IOP in treated (totally regressed) uveal melanoma (both anterior as well as posterior) did not find a greater risk of local or extraocular recurrence after a median follow up of 2 years. [42] However, further analysis with larger sample size and longer follow up duration is warranted, and caution needs to be exercised while deciding regarding treatment options for control of IOP in such cases.

Choroidal metastasis represents around 90% of total uveal metastasis. Breast, lung, kidney, gastrointestinal tract, and cutaneous melanoma are the leading sites of primary malignancies. [32] (**Figure 4A**) Secondary glaucoma has been reported in 1% of cases with mechanism being angle closure and iris neovascularization. [24] Plaque brachytherapy, photodynamic therapy, EBRT, systemic chemotherapy or enucleation are available management options for choroidal metastasis. Secondary glaucoma can be managed like choroidal melanoma related secondary glaucoma described above.

Choroidal hemangioma is a benign vascular hamartoma presenting as an orange- red mass showing hyperreflectivity on ultrasound B scan. It can be solitary or 'circumscribed' or 'diffuse'. (**Figure 4C**) Circumscribed tumors cause secondary glaucoma in 1% of cases of due to angle closure and iris neovascularization due to total retinal detachment. [43] Diffuse choroidal hemangiomas appear as diffuse choroidal thickening and have been said to give 'tomato-catsup' fundus impression. They are known to be associated with Sturge–Weber syndrome (SWS). The raised IOP found in SWS has been found to be due to developmental anomalies of anterior chamber and raised episcleral venous pressure and often unrelated to diffuse choroidal hemangioma. Total retinal detachment and iris neovascularization can rarely cause secondary glaucoma in such eyes. [3] Management options include plaque brachytherapy (for circumscribed as well as to the nodular component if any, in the diffuse variant), photodynamic therapy or EBRT to halt tumor growth and control

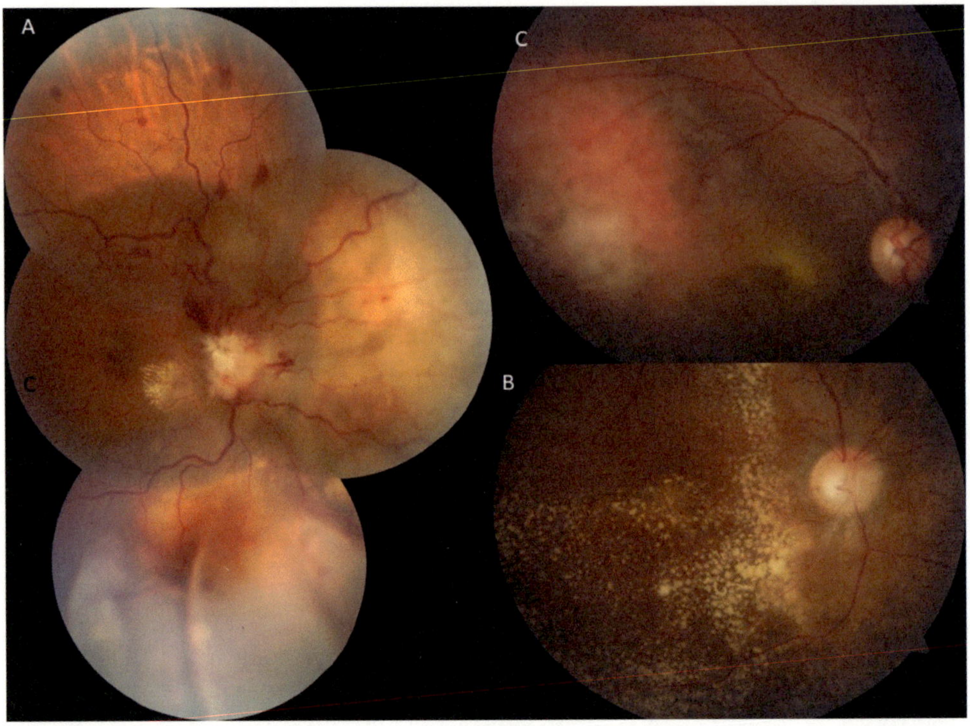

Figure 4.
(A) Montage fundus image of choroidal metastasis from breast carcinoma presenting with multifocal lesions, retinal detachment and raised IOP. (B) Posterior pole fundus image of the same patient (A) showing radiation retinopathy after stereotactic radiotherapy (45Gy) (C) Fundus picture showing circumscribed choroidal hemangioma.

of subretinal fluid to achieve stabilization of visual acuity. [44, 45] For control of IOP, medical management, transscleral cyclophotocoagulation, laser trabeculoplasty, MIGS, filtering or shunting surgeries can be performed.

3.2.2 Retinal tumors

Retinoblastoma is the most common primary intraocular malignancy found in pediatric population with estimated incidence of 1:16,000–18,000 live births. [46] White reflex or 'leukocoria' is the most common presenting feature of retinoblastoma. Presence of an intraocular mass with presence of intratumoral calcifications demonstrated on ultrasound B scan and/or computed tomography (CT) scan is pathognomic of the malignancy. Retinoblastoma can show endophytic, exophytic, diffuse or mixed growth pattern with presence of vitreous seeds. Secondary glaucoma is reported in 17% of cases due to iris neovascularization (5%), angle closure (1%), anterior tumor seeding (<1%) or related to hyphema (<1%). [24] Retinoblastoma is an important differential diagnosis to be considered in any pediatric uveitis or glaucoma. Delayed presentation and often misdiagnosis as either only intraocular inflammation or glaucoma can result in delay in retinoblastoma treatment and greater risk of metastasis with poor prognosis. [1, 2] Neovascular glaucoma is the most common cause of retinoblastoma related raised IOP which is postulated to be mediated by vascular endothelial growth factor (VEGF) produced by necrotic and hypoxic tumor cells. [47] (**Figure 5A** and **B**) Other causes include pupillary block and tumor seeding in anterior chamber angle. Glaucoma in retinoblastoma has been reported to be associated with metastasis related to optic nerve invasion by tumor cells. [48] Presence of secondary glaucoma at presentation has

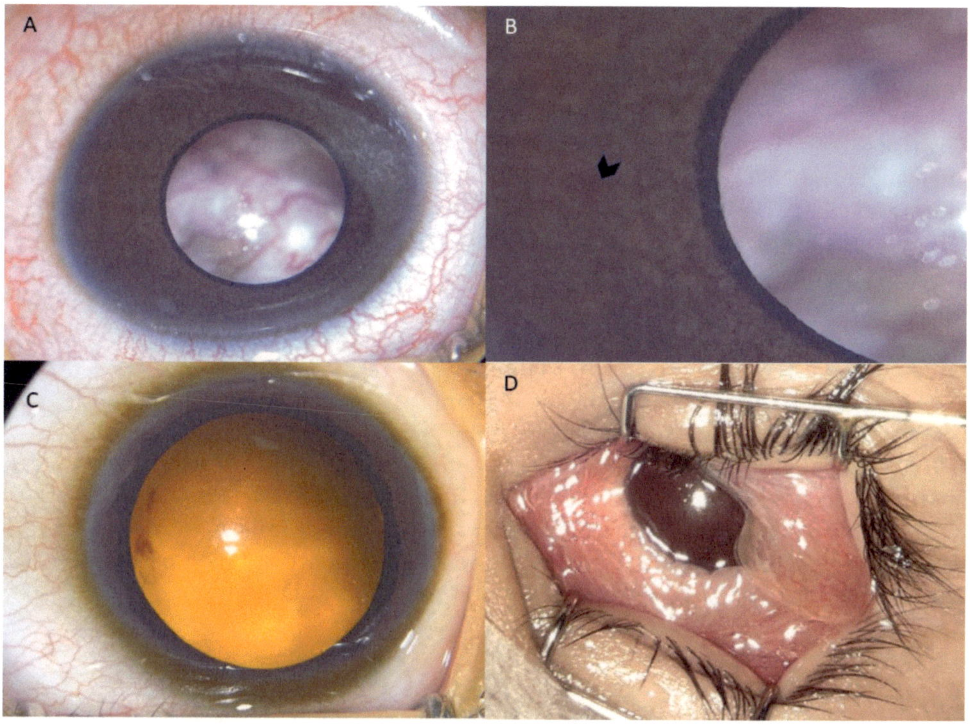

Figure 5.
(A) Group E retinoblastoma presenting with total retinal detachment, iris neovascularization and neovascular glaucoma (clinical risk factors) (B) iris neovascularization (black arrowhead) (C) post chemotherapy tumor necrosis induced vitreous hemorrhage and secondary glaucoma in an eye with retinoblastoma (D) secondary glaucoma and sterile orbital inflammation in a case of retinoblastoma on systemic chemotherapy.

also been identified as a predictor for high-risk histopathological features of retinoblastoma along with prolonged duration of symptoms. [49] Management options for retinoblastoma depend upon age at presentation, laterality as well as grouping and staging of the disease. Modalities used are systemic chemotherapy, intra-arterial chemotherapy, intravitreal chemotherapy, focal lasers, cryotherapy, plaque radiotherapy and Orbital radiotherapy. Non-salvageable and advanced tumor containing eyes need enucleation. The IOP can get controlled as tumor regression is achieved in some eyes precluding any need of separate management for the glaucoma. Medical management, laser trabeculoplasty or cyclophotocoagulation can be tried in selected eyes post-tumor treatment. Any filtering surgeries, shunting procedures, MIGS should be avoided in tumor containing eyes to avoid extra-scleral extension of tumor.

4. Glaucoma associated with management of ocular tumors

Radiotherapy associated glaucoma- Radiotherapy is increasingly being used for management of intraocular tumors as means of eye and vision salvage. The modalities in practice are plaque brachytherapy (Iodine1^{25}, Palladium^{103}, Ruthenium^{106}), proton beam radiotherapy, stereotactic radiotherapy, and external beam radiotherapy. Glaucoma has been reported as one of the common side effects of radiation. The common cause of radiation induced raised IOP is development of iris neovascularization and subsequent neovascular glaucoma (NVG). Pathogenesis of iris neovascularization is multifactorial, including increased release angiogenic factors, anterior and posterior segment ischemia, vascular occlusion, tumor hypoxia, radiation retinopathy (**Figure 4D**) and optic neuropathy. [50] Secondary open angle glaucoma has also been noted in radiotherapy treated tumor eyes apart from NVG. After Iodine ^{125} plaque brachytherapy for uveal melanoma reported Kaplan–Meier estimated risk of secondary open angle glaucoma was 15% at 5 years with higher incidence in earlier postoperative period, whereas estimated incidence of NVG was 13% at 5 years. [51] The risk factors for development of radiation induced secondary glaucoma (both open angle and NVG) include larger tumor size, higher radiation dose, involvement of iris and ciliary body and presence of retinal detachment. [2, 51–53] Proton beam radiotherapy for uveal melanoma has reported NVG rate of 12–31% at 5 years with greater tumor height, older age at presentation and larger tumor diameter as identified risk factors. [54] Anti- VEGF injections as well as pan-retinal photocoagulation for treatment of ocular ischemia or radiation retinopathy might be useful to reduce rate of neovascular glaucoma induced by radiation retinopathy.

Tumor necrosis induced glaucoma- Hemolytic glaucoma (Ghost cell glaucoma) is known to occur in tumor containing eyes with vitreous hemorrhage following systemic chemotherapy or focal tumor treatments. It occurs due to obstruction of trabecular meshwork by red blood cells, their debris and macrophages filled with hemorrhagic components from phagocytosis of vitreous hemorrhage. It can be observed in retinoblastoma eyes showing tumor necrosis as well as vitreous hemorrhage post chemotherapy. (**Figure 5C** and **D**).

4.1 Special considerations

In instances where patient seeks attention with glaucoma as the presenting feature, a detailed work up involving slit lamp examination, dilated fundus examination, gonioscopy of anterior chamber angle, high frequency ultrasound microscopy (UBM) and ultrasound B scan of posterior segment will provide essential clues about diagnosis of possible intraocular tumor. Initial management with IOP lowering medications should be the first line treatment with simultaneous investigation for underlying cause. [2] When the tumor presents with atypical features or causes diagnostic uncertainty, a diagnostic biopsy is warranted in managing such case of secondary glaucoma. [55] Systemic work up with the help of positron emission tomography (PET) CT scan can come in handy while evaluating a case of possible secondary ocular metastasis by highlighting an existing primary malignancy elsewhere.

4.2 Orbital tumor related glaucoma

The orbit is a pyramidal structure limited by bony walls except from anteriorly where it is limited by soft tissue i.e., orbital septum and eyelids. Thus, any instances of increased orbital volume may lead to increased hydrostatic pressure in the orbit. This increased orbital pressure can have a direct effect of IOP by raised hydrostatic pressure around the eye or indirect effect by compression of episcleral and orbital veins raising venous pressure. The episcleral venous system mainly empties into the anterior ciliary and superior ophthalmic veins and eventually draining into cavernous sinus. Thus, any disease process that might affect this drainage pathway due to structural, occlusive, compressive, or destructive pathophysiology can alter and raise IOP causing secondary glaucoma. [4] Focal mass effect due to tumors or swollen extraocular muscles may directly compress the eye globe leading to raised IOP while, vascular changes affecting venous pressure due to compression of episcleral veins or altered arterio-venous flow may also increase IOP indirectly.

Table 1 summarizes the broad etiological classes of orbital tumors leading to rise in IOP and secondary glaucoma. Open angle glaucoma is more common in the listed diseases however, angle closure along with acute angle closure glaucoma has also been reported in variety of pathologies.

Sr. No	Etiology	Example
1	Inflammatory	Non-specific Orbital Inflammatory Disease (NSOID) Thyroid eye disease Orbital granulomatous disease Orbital Foreign Body granuloma Juvenile Xanthogranuloma (orbital histiocytosis)
2	Vascular	Carotid-cavernous fistula (Direct or Indirect) Arterio-venous malformations Orbital varix Cavernous hemangioma Orbital lymphangioma
3	Neoplastic	Orbital osteoma Lymphoproliferative disorders Optic nerve glioma Optic nerve meningioma Neurofibromatosis Lacrimal gland tumors Primary orbital melanoma
4	Secondary	Orbital metastasis (from breast, lung carcinoma) Orbital chloroma (Acute myeloid leukemia) Multiple myeloma Invasive (secondary) ocular melanoma Extra-scleral (orbital) extension of retinoblastoma Invasive (secondary) squamous cell carcinoma
5	Miscellaneous	Orbital amyloidosis Mucopolysaccharidoses Phakomatosis (Neurocutaneous syndromes) Collagen tissue disorders (Lupus erythematosus, Wegener granulomatosis)

Table 1.
Etiologic classification of orbital tumors causing secondary glaucoma.

IOP evaluation should be routinely performed when evaluating a case of suspected orbital tumor or pathology. Gonioscopic examination can provide essential information regarding the status of the anterior chamber angle as well as show evidence of blood in Schlemm's canal as the distinguishing feature of elevated venous pressure. The treatment of primary orbital pathology along with medical management of raised IOP is indicated for control of orbital tumor related secondary glaucoma.

In sum, glaucoma can be associated with various ocular as well as orbital tumors. It may constitute one of the many manifesting clinical features or be the sole presenting feature of these pathologies. Appropriate diagnosis and timely management of these tumors can help eye and vision salvage; however, a misdiagnosis or delayed diagnosis due to initial presentation as secondary glaucoma can lead to catastrophic sequel necessitating enucleation and can pose a greater risk to life. A thorough clinical evaluation, use of ancillary testing and stepwise management can help achieve optimum visual outcome and overall survival in cases with ocular or orbital tumors.

Acknowledgements

The author acknowledges contributions from Dr. Paul T Finger MD, FACS, Director- The New York Eye Cancer Center, New York, USA and Dr. Santosh G Honavar MD, FACS, Director- Ocular Oncology Services, Centre For Sight Hospital, Hyderabad, India for select clinical images used in this chapter.

Author details

Sonal P. Yadav
Dept of Ophthalmic Plastic Surgery, Orbit and Ocular Oncology, PBMA's H V Desai Eye Hospital, Pune, Maharashtra, India

*Address all correspondence to: sonal.yadav@hvdeh.org

References

[1] Shields CL, Shields JA, Shields MB, Augsburger JJ. Prevalence and mechanisms of secondary intraocular pressure elevation in eyes with intraocular tumors. Ophthalmology 1987; 94:839-846

[2] Radcliffe NM, Finger PT. Eye cancer related glaucoma: current concepts. Survey of ophthalmology. 2009 Jan 1; 54(1):47-73.

[3] Camp DA, Yadav P, Dalvin LA, Shields CL. Glaucoma secondary to intraocular tumors: mechanisms and management. Current Opinion in Ophthalmology. 2019 Mar 1;30(2):71-81.

[4] Nassr MA, Morris CL, Netland PA, Karcioglu ZA. Intraocular pressure change in orbital disease. Survey of ophthalmology. 2009 Sep 1;54(5):519-44.

[5] Foulds WS, Lee WR. The significance of glaucoma in the management of melanomas of the anterior segment. Trans Ophthalmol Soc UK. 1983;103 (Pt 1):59—63

[6] Shields CL, Shah SU, Bianciotto CG, Emrich J, Komarnicky L, Shields JA. Iris melanoma management with iodine-125 plaque radiotherapy in 144 patients: impact of melanoma-related glaucoma on outcomes. Ophthalmology. 2013 Jan 1;120(1):55-61.

[7] Demirci H, Mashayekhi A, Shields CL, et al. Iris melanocytoma: clinical features and natural course in 47 cases. Am J Ophthalmol 2005; 139: 468-475.

[8] Fineman MS, Eagle RC Jr, Shields JA, et al. Melanocytomalytic glaucoma in eyes with necrotic iris melanocytoma. Ophthalmology 1998; 105: 492-496.

[9] Cialdini AP, Sahel JA, Jalkh AE, et al. Malignant transformation of an iris melanocytoma. A case report. Graefes Arch Clin Exp Ophthalmol. 1989;227(4):348—5

[10] Finger PT, Latkany P, Kurli M, *et al* The Finger iridectomy technique: small incision biopsy of anterior segment tumours *British Journal of Ophthalmology* 2005;**89:**946-949.

[11] Fountain TR, Goldberg MF, Green WR. Glaucoma and a melanocytic iris lesion in an-18-year-old. In: Schachat AP, editor. Current practice in ophthalmology. St. Louis: Mosby, 1992:371-380.

[12] Chen MJ, Liu JL, Li WY, et al. Diode laser transscleral cyclophotocoagulation in the treatment of refractory glaucoma with iris melanocytoma. J Chin Med Assoc 2008; 71:546-548.

[13] Shields JA, Annesley WH Jr, Spaeth GL. Necrotic melanocytoma of iris with secondary glaucoma. Am J Ophthalmol 1977;84:826-829.

[14] Shields CL, Shields JA, Materin M, et al. Iris melanoma: risk factors for metastasis in 169 consecutive patients. Ophthalmology. 2001;108:172-178.

[15] Khan S, Finger PT, Yu GP, et al. Clinical and pathologic characteristics of biopsy-proven iris melanoma: a multicenter international study. Arch Ophthalmol. 2012;130(1): 57-64.

[16] Singh AD, Turell ME, Topham AK. Uveal melanoma: trends in incidence, treatment, and survival. Ophthalmology 2011; 118:1881-1885.

[17] Demirci H, Shields CL, Shields JA, et al. Ring melanoma of the anterior chamber angle: a report of fourteen cases. Am J Ophthalmol 2001; 132: 336-342.

[18] Van Buskirk EM, Leure-duPree AE. Pathophysiology and electron microscopy of melanomalytic glaucoma. Am J Ophthalmol. 1987;85(2):160—6

[19] Demirci H, Shields CL, Shields JA, et al. Diffuse iris melanoma: a report of 25 cases. Ophthalmology 2002; 109: 1553-1560.

[20] Shields CL, Kaliki S, Furuta M, et al. Iris melanoma features and prognosis in children and adults in 317 patients. J Am Assoc Pediatr Ophthalmol Strabismus 2012;16:10-6.

[21] Chaugule SS, Finger PT. Regression patterns of iris melanoma after palladium-103 (103Pd) plaque brachytherapy. Ophthalmology. 2017 Jul 1;124(7):1023-30.

[22] Girkin CA, Goldberg I, Mansberger SL, et al. Management of iris melanoma with secondary glaucoma. J Glaucoma 2002; 11:71-74.

[23] Bianciotto C, Shields CL, Kang B, Shields JA. Treatment of iris melanoma and secondary neovascular glaucoma using bevacizumab and plaque radiotherapy. Arch Ophthalmol 2008; 126:578-579.

[24] Shields JA, Shields CL, Shields MB. Glaucoma associated with intraocular tumors. In: Ritch R, Shields MB, Krupin T, editors. The glaucomas, 2nd ed. St. Louis, MO: Mosby; 1996. pp. 1131-1139.

[25] Demirci H, Shields CL, Shields JA, Honavar SG, Eagle RC Jr: Ring melanoma of the ciliary body: report on twenty-three patients. Retina 2002;22:698-706; quiz 852-853

[26] Gunduz K, Shields CL, Shields JA, et al. Plaque radiotherapy of uveal melanoma with predominant ciliary body involvement. Arch Ophthalmol 1999;117:170-177

[27] Zhao M, Mu Y, Dang Y, Zhu Y. Secondary glaucoma as initial manifestation of ring melanoma: a case report and review of the literature. Int J Clin Exp Pathol 2014; 7:8163-8169.

[28] Lee V, Cree IA, Hungerford JL. Ring melanoma—a rare cause of refractory glaucoma. Br J Ophthalmol 1999; 83:194-198.

[29] Kaliki S, Eagle RC, Grossniklaus H, et al. Inadvertent implantation of aqueous tube shunts in glaucomatous eyes with unrecognized intraocular neoplasms: report of 5 cases. JAMA Ophthalmol 2013; 131:925-928.

[30] Gragoudas ES, Lane AM, Munzenrider J, Egan KM, Li W: Long-term risk of local failure after proton therapy for choroidal/ciliary body melanoma. Trans Am Ophthalmol Soc 2002; 100: 43-48; discussion 48-49.

[31] Pasternak S, Erwenne CM, Nicolela MT: Subconjunctival spread of ciliary body melanoma after glaucoma filtering surgery: a clinicopathological case report. Can J Ophthalmol 2005; 40: 69-71

[32] Shields CL, Welch RJ, Malik K, et al. Uveal metastasis: clinical features and survival outcomes of 2214 tumors in 1111 patients based on primary tumor origin. Middle East Afr J Ophthalmol 2018; 25:81-90.

[33] Shields CL, Kaliki S, Crabtree GS, et al. Iris metastasis from systemic cancer in 104 patients: the 2014 Jerry A. Shields Lecture. Cornea 2015; 34:42-48.

[34] Seidman CJ, Finger PT, Silverman JS, Oratz R. Intravitreal bevacizumab in the management of breast cancer iris metastasis. Retin Cases Brief Rep 2017; 11:47-50

[35] Aydin A, Tezel TH. Use of intravitreal bevacizumab for the treatment of secondary glaucoma caused by metastatic iris tumor. J Glaucoma 2018; 27:e113–e116.

[36] Chan CC, Buggage RR, Nussenblatt RB. Intraocular lymphoma. Curr Opin Ophthalmol. 2002;13(6): 411—8

[37] Gauthier AC, Nguyen A, Munday WR, et al. Anterior chamber non-Hodgkin lymphoma of the iris masquerading as uveitis-glaucoma-hyphema syndrome. Ocul Oncol Pathol 2016; 2:230-233

[38] Bawankar P, Das D, Bhattacharjee H, et al. Systemic diffuse large B-cell lymphoma masquerading as neovascular glaucoma. Indian J Ophthalmol 2018; 66:317-319.

[39] Matano S, Ohta T, Nakamura S, et al. Leukemic hypopyon in acute myelogenous leukemia. Ann Hematol. 2000;79(8): 455—8

[40] Othman IS, Assem M, Zaki IM. Secondary glaucoma as initial manifestation of uveal melanoma. Saudi Journal of Ophthalmology. 2013 Jul 1;27(3):203-8.

[41] Vempuluru VS, Jakati S, Krishnamurthy R, Senthil S, Kaliki S. Glaucoma as the presenting sign of intraocular tumors: beware of the masquerading sign. International ophthalmology. 2020 Mar 20:1-7.

[42] Fatehi N, McCannel T, A, Giaconi J, Caprioli J, Law S, K, Nouri-Mahdavi K: Outcomes of Glaucoma Drainage Device Surgery in Eyes with Treated Uveal Melanoma. Ocul Oncol Pathol 2019;5:20-27. doi: 10.1159/000488056

[43] Shields CL, Honavar SG, Shields JA, et al. Circumscribed choroidal hemangioma: clinical manifestations and factors predictive of visual outcome in 200 cases. Ophthalmology 2001; 108:2237-2248.

[44] Finger PT. Radiation therapy for exudative choroidal hemangioma. Indian journal of ophthalmology. 2019 May;67(5):579.

[45] Sen M, Honavar SG. Circumscribed choroidal hemangioma: An overview of clinical manifestation, diagnosis and management. Indian journal of ophthalmology. 2019 Dec;67(12):1965.

[46] Kivela T. The epidemiological challenge of the most frequent eye cancer: retinoblastoma, an issue of birth and death. Br J Ophthalmol 2009; 93:1129-1131.

[47] Kvanta A, Steen B, Seregard S. Expression of vascular endothelial growth factor (VEGF) in retinoblastoma but not in posterior uveal melanoma. Exp Eye Res. 1996;63(5): 511—8

[48] Shields CL, Shields JA, Baez K, et al: Optic nerve invasion of retinoblastoma. Metastatic potential and clinical risk factors. Cancer 73:692-8, 1994

[49] Kaliki S, Srinivasan V, Gupta A, Mishra DK, Naik MN. Clinical features predictive of high-risk retinoblastoma in 403 Asian Indian patients: a case-control study. Ophthalmology. 2015 Jun 1;122(6):1165-72.

[50] Sivak-Callcott JA, O'Day DM, Gass JD, et al. Evidence-based recommendations for the diagnosis and treatment of NVG. Ophthalmology 2001;108:1767-1776

[51] Kim EA, Salazar D, McCannel CA, Kamrava M, Demanes DJ, Lamb J, Caprioli J, McCannel TA. Glaucoma After Iodine-125 Brachytherapy for Uveal Melanoma: Incidence and Risk Factors. Journal of glaucoma. 2020 Jan 8;29(1):1-0.

[52] Finger PT. Plaque radiation therapy for malignant melanomaof the iris and ciliary body. Am J Ophthalmol. 2001; 132(3):328—35

[53] Finger PT. Radiation therapy for choroidal melanoma. Surv Ophthalmol. 1997;42(3):215—32

[54] Mishra KK, Daftari IK, Weinberg V, Cole T, Quivey JM, Castro JR, Phillips TL, Char DH. Risk factors for neovascular

glaucoma after proton beam therapy of uveal melanoma: a detailed analysis of tumor and dose–volume parameters. International Journal of Radiation Oncology* Biology* Physics. 2013 Oct 1;87(2):330-6.

[55] Gombos D.S. (2019) Localised Therapy and Biopsies of Intraocular Tumors. In: Chaugule S., Honavar S., Finger P. (eds) Surgical Ophthalmic Oncology. Springer, Cham.